Pre-GED Mathematics and Problem-Solving Skills

BOOK 2

ROBERT MITCHELL

Consultants and Field Testers
Sr. Kathleen Bahlinger, C.S.J.
Sr. Lory Schaff, C.S.J.
St. Paul Adult Learning Center
Baton Rouge, LA

Project Editor
Ellen Carley Frechette

CONTEMPORARY
BOOKS, INC.
CHICAGO • NEW YORK

Library of Congress Cataloging-in-Publication Data
(Revised for book 2)

Mitchell, Robert P.
   Contemporary's pre-GED mathematics & problem-solving
skills.

   (Contemporary's pre-GED series)
   1. Mathematics—1961-   2. Problem solving.
I. Contemporary Books, inc.   II. Contemporary's pre-GED
mathematics and problem-solving skills.   III. Title.
IV. Series.
QA107.M57   1987   510   86-32989
ISBN 0-8092-5150-7 (v. 2)

Material in this book also appears in Contemporary's
*Breakthroughs in Mathematics and Problem Solving
Skills, Book 2.* Copyright © 1989 by Robert P. Mitchell.

Published by Contemporary Books, Inc.
180 North Michigan Avenue, Chicago, Illinois 60601
Manufactured in the United States of America
International Standard Book Number: 0-8092-5150-7

Published simultaneously in Canada by
Fitzhenry & Whiteside
91 Granton Drive
Richmond Hill, Ontario L4B 2N5
Canada

Editorial
Ann Upperco
Sarah Schmidt
Christine M. Benton

Editorial Director
Caren Van Slyke

Production Editor
Patricia Reid

Illustrator
Rosemary Morrissey-Herzberg

Art Director
Georgene G. Sainati

Art & Production
Princess Louise El
Arvid Carlson
Lois Koehler
Jan Geist

Typography
J•B Typesetting

Cover photo © Image Bank

Contents

To The Student

Welcome to *Pre-GED Mathematics and Problem-Solving Skills: Book 2.*

Pre-GED Math Book 2 is designed to give you a wide range of basic arithmetic computation and word-problem skills. These are the core of skills necessary to pass the GED Test. They are also the skills that prepare you for work in algebra and geometry.

Pre-GED Math Book 2 is divided into six chapters. Chapters 3, 4 and 5 are followed by a "Skills Review" that you can use to check your progress in the chapter.

Chapter 1 is a special chapter that introduces you to a variety of word-problem skills. These skills will be used throughout your study of this workbook. To read Chapter 1, you need only be familiar with whole number (and money) computation skills.*

Chapters 2 through 5 introduce new computation skills. These skills involve decimals, fractions, and percents. As you learn these new skills, you will use them to further develop your ability to confidently solve word problems.

Chapter 6, the last chapter of Book 2, is called "Special Topics in Mathematics." The first topic, "Approximation," is a special word-problem skill. The remaining topics, "Measurement," "Simple Interest," "Data Analysis," and "Probability," are important applications of the skills you are learning. Because these topics are important in everyday life, they are included as topics on the GED Test. You'll want to study Chapter 6 very carefully.

Following Chapter 1 is a 4-page "Overview of Computation Skills." This overview will let you know which computation skills you are already familiar with, and which you need to study especially carefully in Chapters 2 through 5. Following Chapter 6 is a word-problem posttest that covers the range of skills discussed in Book 2. You can use this posttest to help identify any skills you may need more practice with.

To get the most out of your work, do each problem carefully. Check your answers to make sure you are working accurately. **An answer key is provided at the back of the book.**

*If you feel that you need more work with adding, subtracting, multiplying, or dividing whole numbers and money amounts, see our other book *Pre-GED Mathematics and Problem-Solving Skills: Book 1.*

1
Word-Problem Skills

What Is a Word Problem?

A word problem is a short story that asks a question. The reader is given numerical information and is asked to find a numerical answer. Example:

Last year the Zarate family paid $315 each month in rent. This year their rent was increased by $15 per month. What is their new monthly rent?

The Importance of Word Problems

Word problems are a very important part of your study of mathematics. Because we speak in words, we almost always express daily life math problems as word problems. Then, to solve a word problem, we replace words with numbers that are to be added, subtracted, multiplied, or divided.

You'll be interested to know that word problems are now the main type of test question that appears on most educational and vocational tests. Included are GED, college entrance, civil service, and military tests. Thus, strengthening word-problem skills is especially important for students who are preparing for a math test in the near future.

Steps in Solving a Word Problem

You may be able to solve many word problems "in your head." This is called **math intuition**. Math intuition may enable you to "see" how to work a problem without really thinking about it. This intuition often works well for short problems containing small whole numbers. For more difficult problems, most people find it useful to use a step-by-step problem-solving approach.

In this chapter, we'll review the 5-step approach to solving word problems. Learning these steps will help organize your thinking and will help strengthen your intuition for all problems. It is important to always read a problem carefully; more than once is helpful. Then, follow these 5 steps:

Step 1. Determine what the question asks you to find.
Step 2. Decide what information is necessary in order to answer the question.
Step 3. Choose the arithmetic operation you wish to use.
Step 4. Solve the problem by doing the math and answering the question.
Step 5. Read the question again and see if your answer makes sense.

Example: Ellen bought hair shampoo for $2.89, hair rinse for $1.98, and a new comb for $1.29. She paid the clerk with a ten-dollar bill. **How much did these three items cost her altogether?**

Step 1. **Understand the question.**

In the example above, the question is in bold type. Reading the question, you want to make sure you can identify what you are asked to find.

asked to find: total cost of three items

Step 2. **Find necessary information.**

Necessary information includes only those **numbers** and **labels** needed to answer a specific question. A label is a symbol or word that tells what the numbers refer to. In the example above, the label is the dollar sign $. The necessary information is the three prices.

necessary information: $2.89, $1.98, $1.29

Step 3. **Choose an arithmetic operation.**

Ask yourself, "To find the total cost, do I add, subtract, multiply, or divide?" To find this total, you add.

operation to use: addition

Step 4. **Solve the problem.**

add the three prices: $2.89
1.98
1.29
Answer: $6.16

Step 5. **Check to see if the answer makes sense.**

Does the answer $6.16 seem about right? Yes, it does. If we had written an answer of $61.60 or $0.616, we would know that a mistake had been made. In that case, we would do the addition again. Or, we may need to go back to step 3 and choose another operation.

On the next few pages, you'll have a chance to practice each of these steps.

Understanding the Question

A word problem may consist of one or more sentences. In most problems, the final sentence is a question. But, in some problems, the question takes the form of an instruction that tells you to compute a certain quantity. **In each type of problem, the first step is to identify the question (or instruction) and to understand what it asks you to find.**

Example 1 contains only one sentence. This sentence is a question that also gives numerical information.

Example 1: **What is the total price of 4 gallons of milk if each gallon costs $1.79?**

asked to find: cost of 4 gallons of milk

Example 2 contains several sentences. The sentence in bold type asks a question, while the other sentences contain numerical information.

Example 2: Fred and three of his friends agreed to evenly split the cost of renting a VCR and 2 movies. The one-day rental rate for the three items is $9.60. **How much is Fred's share?**

asked to find: Fred's share of rental cost

In Example 3 you must **interpret** (figure out the meaning of) what the question asks you to find.

Example: 3: During December it rained on 5 days, hailed on 2 days, and snowed on 7 days. **How many December days had precipitation of some kind?**

asked to find: number of days with precipitation

Although the word *precipitation* appears only in the question, you interpret it to include rain, hail, and snow.

In each problem below, underline the question or instruction. Then, circle the words within the parentheses that best identify what you are asked to find. **Do not solve these problems.**

1. June pays $278.00 each month in rent. Compute the total amount of rent she pays in a year's time.

   (weekly rent, monthly rent, yearly rent)

2. As a janitor, Charlie must clean 16 classrooms each night. If he works an 8-hour shift, how many rooms does Charlie clean each hour?

   (time to clean each room, number of rooms each hour, time to clean 16 rooms)

3. At a garage sale, Jenny bought a table for $15.00, a chair for $10.00, and a lawn mower for $25.50. What amount did Jenny spend at the garage sale?

(cost of furniture, total money spent, number of items bought)

4. In the Swanson household, the television is "on" an average of 21 hours each week. During an average month, about how many hours is the Swansons' TV on?

(TV hours per day, TV hours per week, TV hours per month)

5. While in beauty school, Judy is going to split the cost of household expenses with 2 roommates. These expenses will average about $765 each month. Knowing this, compute Judy's total household expenses for the 9-month school year.

(monthly expenses, nine month expenses, yearly expenses)

6. To tile his shop floor, Vaughn needs 342 tiles. By buying an unopened box of 400 tiles, he got a special price. Instead of paying 79¢ for each tile, he only had to pay 59¢. How much did the tiles cost Vaughn?

(cost of 400 tiles, cost of 342 tiles, amount saved on purchase)

7. There were 8 women and 12 men enrolled in Maxine's class as of Monday. On Tuesday 3 more women enrolled. Determine how many students are in Maxine's class now.

(number of women, number of men, number of women and men)

8. Out of her monthly paycheck, Milly pays $123 in federal income tax, $64 in state income tax, and $8.35 in city income tax. How much does Milly pay in state and city taxes each month?

(city and state taxes, total tax expense, federal and state taxes)

9. Orsen has a handful of change in his pocket. He has three quarters, four dimes, and two nickels. How many coins does he have in all?

(total value of change, number of coins, number of cents)

10. During April, Diane paid $175 for rent, $25 for sewer and water, $95 for electricity, and $135 for a car payment. Not counting the car payment, what was the sum of Diane's April payments?

(total April payments; April payments for rent, sewer, and water; April payments for rent, sewer, water, and electricity)

Finding Necessary Information

Word problems often contain more information than is needed to answer the question. Therefore, it is important that you be able to select only the information you need. This information is called **necessary information**.

> *Necessary information* includes *only* those numbers and labels needed to answer a specific question.

> *Given information* includes *all* of the numbers and labels that appear in a problem. *Extra information* is information that appears in a problem but is not needed to answer the question.

In choosing necessary information, ask yourself, "What information do I need to answer this question?" Then, reread the problem carefully and select only those numbers and labels that you need.

Example: While shopping the New Year's sale, Maria bought a dress for $24.95, a skirt for $13.50, and a blouse for $7.99. She paid the clerk with a check for $50.00. What was the total price of the clothes she bought?

asked to find: total price of clothes

given information: $24.95, $13.50, $7.99, $50.00

necessary information: $24.95, $13.50, $7.99

Notice that the question asks only for the total price of the clothes. Nothing is asked about how much she paid the clerk or about how much change she got back. Thus, $50.00 is extra information, not necessary information.

In each problem below, underline the question or instruction. Then write the necessary information on the line beneath the question. **Do not solve these problems.**

1. At Gina's Market, Fran bought cereal for $2.89, milk for $1.79, grapes for $1.43, toothpaste for 99¢, and paper towels for 89¢. What amount did Fran pay for the three food items?

   _____

2. On the first day of Sal's swimming class, 6 girls and 5 boys showed up. Three more girls and 4 more boys joined before the end of the week. By week's end, how many girls were in Sal's class?

   _____

3. For the church picnic, Peggy bought 500 paper cups. Although she expected 275 people to attend, 350 showed up. At 4¢ per cup, how much did Peggy spend on cups?

_____

4. Amy shopped at Penney's weekend clothing sale. Among other things, she bought a coat for $28.45, a scarf for $2.89, and a blouse for $7.98. After paying with a check for $50.00, Amy received $3.61 in change. Compute the total amount that Amy spent at Penney's sale.

_____

In some problems, necessary information is not directly given. You may have to identify needed numbers and labels by interpreting words in the problem or by looking at a list or at a drawing.

5. Chicago time is 1 hour earlier than New York time and 2 hours later than Seattle time. What time is it in Chicago when it is 3:00 P.M. in Seattle?

_____

6. Al's Tune-Up Shop ran a weekend special as shown at right. Richard had Al change his oil, put in a new oil filter, and rotate his tires. He also had Al replace two burned-out taillights at a cost of $1.25 each. Figure out the total amount that Richard spent at Al's.

_____

Special Weekend Rates

Oil & Filter Change	$9.95
Tires Rotated	$4.00
Car Lubricated	$5.75
Wash & Wax	$19.95

7. Henry is going to drive from Lakeridge to Duggin by going through Springfield. Looking at the map at right, which route—through Orny or Likely—is shorter?

_____

Lakeridge
243   381
278   Orny
312   Springfield
Likely
326   Duggin

7

Choosing an Arithmetic Operation

Students often ask, "How can I be sure when to add, subtract, multiply, or divide?" There is no easy answer to this question. In many problems, you will simply know what to do. In problems where you're not sure, you look for clues.

Addition and Subtraction Problems

One clue to solving a problem is often found in certain words of the problem. These clue words are called **key words**. Key words may suggest adding or subtracting.

Example 1: Ira hiked 19 miles on Saturday. On Sunday he hiked 12 more miles. How many *total* miles did Ira hike on Saturday *and* Sunday?

Add: 19
+ 12

Answer: **31 miles**

key words: total, and

The key words *total* and *and* suggest combining things to find a total. Example 1 is solved by adding.

Example 2: Ira hiked 19 miles on Saturday. On Sunday he hiked 12 more miles. How much *farther* did Ira hike on Saturday *than* on Sunday?

Subtract: 19
− 12

Answer: **7 miles**

key words: farther, than

The key words *farther* and *than* suggest finding the difference between two distances. Example 2 is solved by subtracting.

A second clue is often obtained by drawing a picture or diagram. A drawing is especially helpful in problems where key words are not easy to identify. In Example 3, there is no single key word that correctly suggests adding.

Example 3: Between noon and 7:00 P.M. on Tuesday, the temperature dropped by 26°. At 7:00 P.M., the temperature was 43° F. What had the temperature been at noon?

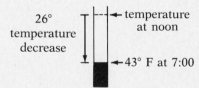

As the picture helps show, the noon temperature is *greater than* either of the two numbers given in the problem. Thus, to find the noon temperature, you add 26° and 43°. You do not subtract.

Add:	43° F
	+26° F
Answer:	**69° F**

Drawing a diagram helps because it causes you to think carefully about the numbers in a problem. You must ask yourself, **"Is the number I'm trying to find smaller or larger than the other numbers in the problem?"** Answering this question is the key to solving all addition and subtraction problems.

Each problem below contains one main key word that suggests either addition or subtraction. In each problem, underline this key word. Then, on the line below each problem, write *add* if the key word suggests adding or *subtract* if the key word suggests subtracting. Finally, solve each problem and check your answers.

1. While on a special diet, Francis ate 1,200 calories fewer each day than her regular 2,600 calories. How many calories each day did this diet allow Francis to eat?

2. Starting in November, the Jones family is getting a rent increase of $25 per month. If they are now paying $295 per month, what will be their new monthly rent?

3. Jack's car weighs 2,975 pounds. His trailer weighs 475 pounds when empty. When full, the trailer can carry a load of 1,225 pounds. What is the combined weight of the car and loaded trailer?

4. While shopping at Save-All, Lita used food stamps to buy vegetables for $3.47, fruit for $4.78, and salad dressing for $1.79. To buy these items, what total value of food stamps will Lita need?

5. During the storm, flood water on Main Street rose to 5 feet by Sunday evening. Twelve hours later, the level had decreased by 4 feet. How high was the flood water on Main Street by Monday morning?

6. At 3:00 P.M. the temperature in Elsy was 98° F. By 7:00 that evening, the temperature had fallen by 39° F. What was the temperature in Elsy at 7:00 P.M.?

In problems 7, 8, and 9, use the drawing to help you decide whether to add or to subtract. Then solve each problem and check your answer.

7. Becky Robinson borrowed $1,200 to buy new furniture. So far, she has paid back $850. Not counting any interest charges, how much does Becky still owe?

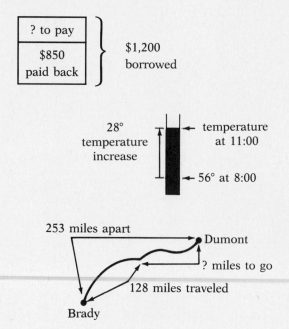

8. Between 8:00 A.M. and 11:00 A.M. on Friday, the temperature rose 28° F. What was the temperature at 11:00 if the temperature at 8:00 was 56° F?

9. The distance between Brady and Dumont is 253 miles. If Yoshi has traveled 128 miles of this distance, what distance does she still have to drive?

In problems 10, 11, and 12, complete each drawing by filling in the blank lines with numbers from the problems. Then use your completed drawing as a guide and solve each problem.

10. James Olsen paid $236 more in taxes in 1986 than he paid in 1985. If he paid $873 in taxes in 1985, how much did he pay in taxes in 1986?

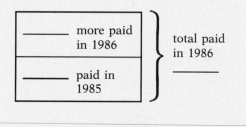

11. Between 7:00 A.M. and noon on Saturday, the temperature rose 34° F. What was the temperature at 7:00 if the noon temperature was 89° F?

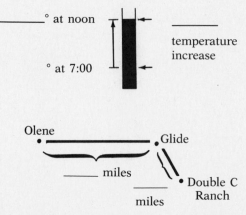

12. The distance between Olene and Glide is 47 miles. From Glide it is only 19 miles to the Double C Ranch. What is the distance between Olene and the ranch if you first go through Glide?

Multiplication and Division Problems

As the following examples show, key words may also appear in multiplication and division problems.

Example 1: During one rush hour, Kim sold 144 medium-size cups of orange pop. If each cup holds 12 ounces, how many ounces of pop did she sell *altogether*?

Multiply:
$$\begin{array}{r} 144 \\ \times\ 12 \\ \hline 288 \\ 1\,44 \\ \hline \end{array}$$

Answer: **1,728 ounces**

key word: altogether

The key word *altogether* suggests combining (adding or multiplying) things to find a whole amount. Example 1 is most easily solved by multiplying.

Example 2: During one rush hour, Kim sold one full container of grape drink. To sell this drink, she used 96 large-size cups. If the full container held 1,536 ounces, how much on the *average* did *each* cup hold?

Divide:
$$\begin{array}{r} 16 \\ 96\overline{)1536} \\ 96 \\ \hline 576 \\ 576 \\ \hline 0 \end{array}$$

Answer: **16 ounces**

key words: average, each

The key words *average* and *each* suggest finding part of a whole amount. Example 2 is solved by dividing.

For problems in which you're unable to find key words, you may find it helpful to write a **solution sentence**. A solution sentence uses words to briefly state a problem's solution. Solution sentences are especially helpful in multiplication and division problems because they help you form a mental picture of how to solve the problem.

Example 3: Garbage service in Greenville costs $102 each year. At this rate, what is the monthly charge for garbage collection?

Here is a solution sentence for this problem:

*monthly charge = yearly charge **divided by** number of months in a year*

Replacing words with numbers, we obtain:

monthly charge = $102 ÷ 12 = $8.50

Answer: **$8.50 each month**

Each problem below contains one or two key words that suggest either multiplication or division. In each problem, underline any key word you can find. Then, on the line below each problem, write *multiply* if the key word(s) suggests multiplying or *divide* if the key word(s) suggests dividing. Finally, solve each problem and check your answers.

13. Sally and her two sisters agreed to split the cost of lunch. If the bill for sandwiches, drinks, and desserts amounted to $15.75, how much is Sally's share?

_____

14. The regular price of a 12-ounce container of frozen orange juice is twice the sale price. What is the regular price if the sale price is 59¢?

_____

15. As part of her fitness program, Barbara takes a 5-mile jog four times each week. In an average week, how many miles does Barbara jog in all?

_____

16. Stan has to deliver 48 appliances during a 6-hour period on Saturday morning. To finish in time, how many appliances does Stan need to deliver every hour?

_____

17. How many 9-inch pieces can be cut from a board that measures 72 inches in length?

_____

18. Paying $1.29 per gallon, how much would 18 gallons of gas cost altogether?

_____

Problems 19, 20, and 21 are each followed by a **solution sentence**. Complete each sentence by writing either the word *multiplied* or *divided* on the blank line. Then use the completed solution sentence as a guide to solve each problem.

19. The Johnson family drinks 10 gallons of milk each month. Remembering there are 4 quarts in 1 gallon, how many quarts of milk each month is this?

quarts each month = gallons each month _____ by the number of quarts per gallon

20. Out of each monthly paycheck, Manny puts $85 in his savings account. At this rate, how many months will it take Manny to save the $765 he needs to be able to buy a new stereo system?

number of months = amount to be saved _____ by the amount saved each month

21. Wendy, a truck driver, must haul 736 crates to a warehouse across town. Usually she would not haul more than 85 in one trip. However, this is a rush job. If she has time to make only 8 trips, how many crates must Wendy haul on each trip?

number of crates = total number to be moved _____ *by the number of trips*

In problems 22, 23, and 24, use short word phrases to complete each solution sentence. Then solve these problems.

22. Robbie exercises for 5 hours every week. Remembering there are 52 weeks in a year, how many hours does Robbie exercise during a year's time?

hours of exercise each year = **hours per week** *multiplied by* _____

23. By saving $125 each month, Julie has managed to save $1,750. How many months has it taken Julie to save this much?

number of months needed = _____ *divided by* _____

24. In the United States, an average car owner drives about 1,250 miles each month. At this rate, how many miles does an average car owner put on a car in a year's time?

mileage each year = _____ *multiplied by* _____

Checking Your Answer

Key words, drawings, and **solution sentences** often provide clues that help you decide how to solve a word problem. You get one more clue when you **check your answer**.

Checking your answer involves two steps:

1. Checking to see that you did the math correctly.

2. Checking to see that your answer makes sense.

Seeing if your answer makes sense is the most imporant clue you have for knowing if you worked a problem correctly. You ask yourself, "Is my answer about what I expect it should be?" If it isn't, you may have chosen the wrong operation. For example, you may have added when you should have subtracted.

Example: If Jeff received $4.11 in change after paying the clerk $20.00 for a blanket, what was the cost of the blanket?

Suppose you aren't sure whether to solve this problem by adding or by subtracting. One thing you could do is to look at both possible answers:

Solved by adding	*Solved by subtracting*
$20.00	$20.00
+ 4.11	− 4.11
$24.11	$15.89

Of the two answers, your own experience tells you that only $15.89 can be correct. Therefore, the example is a subtraction problem. The addition answer, $24.11, is larger than the $20.00 given to the clerk. $24.11 doesn't make sense as an answer.

When students choose a wrong operation, they most often mix up addition with subtraction, or they mix up multiplication with division. When *you* are unsure about a problem, don't hesitate to solve it in more than one way. **Most often, as shown in the example, only one answer makes sense, and that will be the correct answer.**

To the right of each problem below, circle the answer that makes the most sense. Don't do any computation. Just choose the answer that makes sense to you. Only one answer is correct.

1. Krisa received $31.25 in change after buying a 7-pound roast that cost $13.77. She paid by check. How much did she write the check for?

    a) addition: $45.02
    b) subtraction: $17.48

2. A case of oil is on sale for $9.84. If there are 12 quarts in each case, determine the sale price per quart.

    a) multiplication: $118.08
    b) division: $0.82

3. Four friends agreed to evenly share the cost of driving across the country. If each person's share for gas came to $64, what was the total gas bill?

    a) multiplication: $256
    b) division: $16

4. By 6:30 P.M. the temperature had cooled 29° F below its 3:00 P.M. high. If the temperature at 3:00 was 73° F, what was the temperature at 6:30?

    a) addition: 102°
    b) subtraction: 44°

5. During the weekend sale, the price of a used Ford was reduced by $595. If the price had been $3,795, what was the sale price?

    a) addition: $4,390
    b) subtraction: $3,200

6. A recipe calls for 3 teaspoons of salt to season a meat dish for 6 people. If the cook wanted to prepare the same dish for 18 people, how many teaspoons of salt would be needed?

    a) multiplication: 9
    b) division: 1

7. Cleo's diet was a great success. In only 3 months she lost 17 pounds. Cleo now weighs 146 pounds. What was her weight 3 months ago?

    a) addition: 163
    b) subtraction: 129

8. A youth group divided its membership list into 4 mailing lists. If each mailing list contained 72 names, how many names were on the membership list?

    a) multiplication: 288
    b) division: 18

9. At a price of 24¢ each, how much would 2 dozen bagels cost?

    a) multiplication: $5.76
    b) division: 12¢

10. Ernie paid $780 in property taxes last year. Later he got a $75 refund. How much did Ernie end up paying for property taxes after all?

    a) addition: $855
    b) subtraction: $705

Solving One-Step Word Problems

On these next three pages are one-step word problems for you to solve. A **one-step word problem** is solved by one arithmetic operation. You compute the answer by adding, subtracting, multiplying, or dividing.

While working these problems, remember to read each problem carefully. Make sure you understand what the question asks you to find. Then choose the **necessary information** and decide which arithmetic operation you wish to use. In some problems **key words** may be helpful. In other problems you may want to make a **drawing** or to write a **solution sentence**.

Here are some general guidelines to help you choose the correct operation:

- when combining amounts —————————————————→ add

- when finding the difference between
  two amounts —————————————————→ subtract

- when given one unit of something and
  asked to find several —————————————————→ multiply

- when given a part of something and
  asked to find the total —————————————————→ multiply

- when given an amount for several and
  asked to find an amount for one —————————————————→ divide

- when splitting, cutting, sharing, etc. —————————————————→ divide

After you've solved each problem, check to make sure your answer makes sense. If it doesn't, redo the problem until you're happy with the answer.

Solve each problem below.

1. Work at the Data Corporation is divided into shifts. Ninety-six people work on day shift, 78 work on swing shift, and 67 work on evening shift. How many people in all does the Data Corporation employ?

2. Norma pays $205 in rent each month. What amount of rent does Norma pay in one year's time?

3. As a supervisor, Jan earns $8.67 an hour. Ron is a mail clerk and makes $5.49 an hour. What is the difference between what Jan earns each hour and what Ron earns?

4. Jason and 3 friends agreed to evenly split the cost of renting a camper for the weekend. If the camper rent total was $64.88, what was Jason's share of the cost?

5. A railroad flatcar is carrying 12 new Toyota sport model cars. Each car weighs 2,935 pounds. Compute the total weight carried by this flatcar.

6. After paying for the gallon of milk with a five-dollar bill, Gina was given $2.18 in change. She thinks the clerk made a mistake. How much did he charge her for the milk?

7. Jody's boss wants her to deliver 48 refrigerators on Saturday. On that day she'll have time to make only 6 trips to the warehouse. Assuming she carries the same number each trip, how many refrigerators should Jody pick up on each trip to the warehouse?

8. After working his regular shift last week, Hank worked 7 overtime hours on Saturday. If overtime pay is $9.25 per hour, how much extra money did Hank earn for this Saturday work?

9. On Tuesday it was Debbie's turn to bring snacks for the other kids at school. Debbie's class contains 12 girls and 11 boys. Counting 1 extra for the teacher, how many snacks should Debbie bring to school?

10. In a game of bridge, 52 cards are dealt out equally to 4 players. As the game begins, how many cards are in each player's hand?

Questions 11 through 14 contain extra information. In each, underline the necessary information and then solve the problem.

11. The Wildcat Football Stadium holds 14,500 people. For the first game of the season, 8,684 fans showed up. The second game drew 9,832, and the third game drew 11,459. What was the combined attendance at the first 3 games?

12. At a garage sale, Trudie bought several items. Among other things, she bought a picture for $2.50, a lamp for $5.00, and a sewing machine for $45.00. After she paid with a $100 bill, she got back $31.50 in change. What total amount did Trudie spend at this sale?

13. On the hottest day last summer, the temperature soared to its highest point, 104°, at 4:20 P.M. By 8:00 P.M. the temperature was down to 81°. Finally, by 10:30, the temperature had dropped several more degrees. As shown at the right, how many degrees did the temperature drop between 4:20 and 10:30?

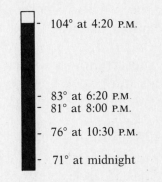

- 104° at 4:20 P.M.

- 83° at 6:20 P.M.
- 81° at 8:00 P.M.

- 76° at 10:30 P.M.

- 71° at midnight

14. At a "White Sale," Penney's had the items shown at right on sale at special prices. How much did Perry spend when he bought the following things: 2 pillow cases, 1 sheet, and 1 bath towel?

White Sale	
Pillow case	$4.49
Sheet	$12.99
Blanket	$29.00
Mattress cover	$26.95
Bath Towel	$6.89

Word problems are often written as a short story followed by two or more questions. Each question requires its own necessary information. Choose this information carefully as you answer each question below.

Questions 15 through 17 refer to the following story.

While shopping at Al's Hardware, Ray bought all 7 items he had written on his shopping list. He was especially pleased to find 3 of those items at sale prices: a new hammer for $7.99, a new hand saw for $13.85, and a 6-pound sack of nails for $5.28.

After paying the clerk $42.00, Ray got back $1.89 in change.

15. How much did Ray spend for the 3 items he bought on sale?

16. How much did Ray pay for each pound of nails?

17. What total amount did Ray spend at Al's Hardware?

Questions 18 through 20 refer to the story below.

Lois borrowed $2,000 from her credit union in order to buy a used car. To repay the loan, Lois can choose to make payments over either 24 months or 36 months.

If she chooses 24 months, Lois must make 24 equal monthly payments of $102.25 each. If she chooses 36 months, Lois must make 36 equal payments of $74.60.

18. What is the difference in months between the two possible loan repayment times?

19. If she chooses the 24-month loan, what total amount will Lois pay to her credit union?

20. If she chooses the 36-month loan, what total amount will Lois pay to her credit union?

Solving Multi-Step Word Problems

So far in this chapter, you have worked with **one-step word problems**. In a one-step problem you either add, subtract, multiply, or divide to compute an answer.

On the next few pages you'll work with **multi-step word problems**. In a multi-step problem you use two or more operations to compute an answer. However, you can solve a multi-step problem by first breaking it down into two or more one-step problems.

To break a multi-step problem into simpler one-step problems, start with a **solution sentence**. To write this sentence, use brief phrases and numbers to state the problem's solution as simply as possible.

Example: Harry bought supplies for his painting business. He bought 24 gallons of paint at a cost of $7.45 each. And he bought 4 paintbrushes at a cost of $3.80 each. How much did these supplies cost Harry?

Here is a solution sentence for this problem:

total cost of supplies = cost of paint plus cost of paintbrushes

Notice that neither the **total cost of paint** nor the **total cost of paintbrushes** is given in the problem as a single number. Because of this, each of these amounts is called *missing information*. Computing missing information is the first step in finding a problem's solution.

In the example, each item of missing information has its own necessary information.

missing information		*necessary information*
total cost of paint	$\longrightarrow$	24, $7.45
total cost of brushes	$\longrightarrow$	4, $3.80

Each value is computed by multiplication:

total cost of paint = $7.45 $\times$ 24
                     = $178.80
total cost of brushes = $3.80 $\times$ 4
                       = $15.20

Now we can place numbers in the solution sentence and compute the answer by addition:

total cost of supplies = $178.80 + $15.20
                        = $194.00

Answer: $194.00

Solve each of the following problems. A solution sentence is written beneath each problem, and the missing information is in bold type. As your first step, compute the value of the missing information.

1. While shopping, Gerald bought chicken for $4.83, a gallon of milk for $1.85, and a loaf of bread for $1.29. If he paid the checker with a ten-dollar bill, how much change should he receive?

   change = $10.00 minus **the total cost of groceries**

2. Ellen baked 87 cookies. She kept 24 cookies at home for her own family, and sent the rest with her daughter Amy to school. If there are 21 students in Amy's class, how many cookies can each child have?

   cookies per child = **cookies taken to school** *divided by 21*

3. At a shirt and sweater sale, David bought 4 shirts for $8.95 each and 3 sweaters for $13.45 each. How much was David charged for this purchase?

   cost of purchase = **total cost of shirts** *plus* **total cost of sweaters**

4. While studying her history book, it takes Sheila 6 minutes to read each page. If she reads history for 90 minutes each day, how many pages can she study during a 5-day period?

   total pages = **pages read per day** *times 5*

5. Ted bought 5 quarts of oil on sale for a price of 83¢ per quart. He also sent in a "5-quart rebate certificate" that returned $1.50 to him because of his purchase. Considering the rebate, how much did the 5 quarts of oil cost Ted?

   cost of oil = **store price of 5 quarts** *minus $1.50*

In problems 6 through 10, complete each solution sentence by using words to identify the missing information. Then, using your solution sentences, solve each problem.

6. Mark and four friends agreed to evenly split the cost of dinner. The bill included $23.40 for the main meal, $6.75 for desserts, and $5.25 for drinks. How much was Mark's share of the dinner bill?

   Mark's share = _total bill_ *divided by 5*

7. Dolores drove 812 miles during her vacation to San Francisco. While on the trip, her car got 28 miles for each gallon of gas. If she paid $1.30 per gallon, figure out her total gas expense for this trip.

   gas expense = $1.30 times _____

8. Adam's oil truck was filled with 8,250 gallons of oil when he began deliveries on Monday. By Friday he had made 46 deliveries, averaging 126 gallons per delivery. At this point, how much oil remained in Adam's delivery truck?

oil remaining = 8,250 gallons minus _____

9. Jim's job is to package and ship Wonder Toys. During each week, he packages and ships 35 mailing boxes of these toys. Each mailing box contains 24 separate Wonder Toys. At this rate, how many toys can Jim ship in one year? (1 year = 52 weeks)

toys shipped in one year = _____ *times 52*

10. Rosa wants to color 9 dozen eggs for the Easter egg hunt on Sunday. If she has only 6 colors of dye to use, how many eggs should she dye each color? (1 dozen eggs = 12 eggs)

number of eggs each color = _____ *divided by 6*

Solve problems 11 through 14. In each problem, if you find it helpful, write a solution sentence to help you identify missing information.

11. Eight pizzas were ordered for lunch for the 32 employees of A&D Electronics Company. If each pizza is cut into 12 slices, how many slices would be each employee's share?

12. Stella runs her own diaper-cleaning business. For this work she uses 4 washing machines. Each machine can wash 30 diapers in one load, and one load takes about 20 minutes to complete. How many diapers can Stella clean each hour in each machine?

13. At a Sears Tool Sale, Tony bought a hammer for $4.98, a set of wrenches for $9.95, and a saw for $7.49. If he paid with a check for $30.00, how much change would Tony receive?

14. As a bus driver, Anna drives 6 routes each day on Monday, Wednesday, and Friday. Each Tuesday and Thursday she drives 8 more routes. During her 5-day work week, how many bus routes in all does Anna drive?

Becoming Familiar with Arithmetic Expressions

As you've seen in multi-step word problems, more than one operation may be needed to solve a problem. Here we'll see how arithmetic expressions that contain more than one operation are written and evaluated.

Multiplication and division are always performed before addition and subtraction. Look at these examples:

Arithmetic Expression	Meaning	Computing the Value
$5 \times 4 + 8$	add 8 to the product of 5×4	Compute the product: $5 \times 4 = 20$ Add 8 to 20: $20 + 8 = \mathbf{28}$
$11 - 15 \div 3$	subtract the quotient $15 \div 3$ from 11	Compute the quotient: $15 \div 3 = 5$ Subtract 5 from 11: $11 - 5 = \mathbf{6}$

Next, we'll look at the use of parentheses ( ). **Parentheses are used to indicate a single quantity that is to be multiplied or divided.** Using parentheses, no multiplication sign is needed. Here are two examples:

Arithmetic Expression	Meaning	Computing the Value
$5(9 + 6)$	multiply 5 times the sum of $9 + 6$	Compute the sum: $9 + 6 = 15$ Multiply 15 by 5: $5 \times 15 = \mathbf{75}$
$(15 - 6) \div 3$	divide 3 into the difference of $15 - 6$	Compute the difference: $15 - 6 = 9$ Divide 9 by 3: $9 \div 3 = \mathbf{3}$

When an expression contains parentheses, follow these steps:

Step 1. Do the arithmetic indicated within parentheses first.

Step 2. Starting at the left, do all multiplication and division.

Step 3. Again starting at the left, do all addition and subtraction, adding or subtracting two numbers at a time.

Study the following examples carefully. A bracket ⌐—⌐ indicates the operation performed as you move from one line to the next. Notice that in each step an operation is performed on only two numbers at one time.

Examples:

1. $57 - \underbrace{8 \times 6} + 5$

   $= \underbrace{57 - \quad 48} + 5$

   $= \quad\quad 9 + 5$

   $= 14$

2. $\underbrace{5 \times 7} + \underbrace{72 \div 8}$

   $= \quad 35 + \underbrace{72 \div 8}$

   $= \quad 35 + \quad 9$

   $= 44$

3. $6\underbrace{(13 - 9)} - 14$

   $= \quad \underbrace{6(4)} - 14$

   $= \quad 24 - 14$ [since $6(4)$

   $\quad\quad\quad\quad\quad$ means 6×4]

   $= 10$

Evaluate each expression below. Begin each set of problems by first completing the row of partially worked Skill Builders.

Skill Builders

1. $6 \times 3 + 7$ $39 - 7 \times 5 + 18$ $14 \times 3 - 25 \div 5$

   $= 18 + 7$ $= 39 - 35 + 18$ $= 42 - 25 \div 5$

   $=$ $= \underline{\quad} + 18$ $= 42 - \underline{\quad}$

                          $=$ $=$

---

2. $5 \times 3 + 9$ $8 \times 5 - 17$ $42 - 8 \times 3 + 4$

3. $36 \div 4 + 8$ $6 \times 5 + 7 \times 8$ $12 \times 3 - 45 \div 5$

Skill Builders

4. $6(5 + 2)$ $(32 - 17) \div 3$ $5(7 + 3) - 24 \div 8$

   $= 6(7)$ $= (15) \div 3$ $= 5(10) - 24 \div 8$

   $= 6 \times 7$ $= 15 \div 3$ $= 50 - \underline{\quad}$

   $=$ $=$ $=$

---

5. $8(6 + 3)$ $13(12 - 5) + 3$ $(28 - 4) \div 6$

6. $9(13 - 6) + 5 \times 8$ $36 - (3 + 6 + 4)$ $2(8 + 4) - 49 \div 7$

An arithmetic expression is often used to show the computation steps of a multi-step word problem. **To write this expression, replace the words of the solution sentence with numbers before you compute the value of missing information**. In many problems, there may be more than one correct solution sentence and arithmetic expression.

Example: Cathy earns $5.40 each hour she works on the weekend. If she worked 9 hours last Saturday and 6 hours last Sunday, how much did she earn on those two days?

There are two solution sentences that show how to solve this problem. Each becomes a correct arithmetic expression when words are replaced by numbers.

1. total earned = Saturday's earnings *plus* Sunday's earnings
                 = ($5.40 × 9) + ($5.40 × 6)

or

2. total earned = pay per hour *times* total hours worked
                 = $5.40(9 + 6)

Each arithmetic expression gives an answer of $81.00.

Learning to write arithmetic expressions to solve word problems is an important skill. You'll use this skill throughout your study of mathematics.

From the choices at right, circle the one expression that will give the correct answer to each problem.

7. Last week, Greg made 8 deliveries each day Monday through Friday. On Saturday he worked until noon and made 3 more deliveries. How many total deliveries did Greg make during the 6 days?

    a) 8 + 3
    b) 5 (8 + 3)
    c) 5 × 8 + 3

8. Shirley planned to divide her penny collection evenly among her 3 children. She had 536 pennies in all. First, she took out 120 pennies she wanted to keep herself. Of the pennies left, how many should she give each child?

    a) 536 ÷ 3 − 120
    b) (536 − 120) ÷ 3
    c) 536 − 120 ÷ 3

9. Jane gave a clerk a twenty-dollar bill to pay for one gallon of paint and one paintbrush. If the paint cost $11.95 and the brush cost $3.45, how much change did she receive?

    a) $20.00 − ($11.95 + $3.45)
    b) ($11.95 + $3.45) − $20.00
    c) $20.00 − $11.95 + $3.45

10. For his cleaning business, Arnie bought 12 gallons of soap at $9.25 per gallon. He also bought a gallon of deodorizer for $17.85. What amount did he pay for these supplies?

a) 12 ($9.25 + $17.85)
b) $9.25 × 12 + $17.85
c) $9.25 (12 + $17.85)

11. On Halloween, Betty bought 4 bags of candy. Each bag contained 75 small wrapped candies. If she expected to see about 60 children, about how many candies should Betty give to each child?

a) (75 × 4) ÷ 60
b) (75 − 60) ÷ 4
c) 4 (75 ÷ 60)

12. George and five friends agreed to evenly split the cost of dinner. They bought a pizza for $13.50, salads for $9.00, and drinks for $3.50. How much did George pay for his share?

a) ($13.50 + $9.00 + $3.50) ÷ 5
b) 6 ÷ ($13.50 + $9.00 + $3.50)
c) ($13.50 + $9.00 + $3.50) ÷ 6

For each problem below, there are two expressions that will give the correct answer. Circle the **two** correct choices to the right of each problem.

13. When fully loaded, Cindy's truck can haul 2,300 pounds of gravel. On Saturday, she hauled 6 full loads. On Sunday, she hauled 5 more full loads. How many pounds of gravel did she haul on these two days?

a) 2,300 (5 + 6)
b) 2,300 × 2
c) 6 + 5(2,300)
d) (2,300 × 5) + (2,300 × 6)
e) 2,300 × 6 + 5

14. Each day he fights fires, Manuel earns $25 more than his regular daily pay of $63. Last week he fought fires on 3 days. How much did he earn during the 5 day week?

a) 5 ($63 + $25)
b) $63 × 5 + $25 × 3
c) $63 × 2 + $25 × 3
d) $63 × 5 + $25 × 2
e) 3 ($63 + $25) + 2 × $63

15. Sam, Frank, and Lois bought a boat. The sale price was $5,600. They also received a $1,200 rebate a month later. If they split all costs evenly, how much did each one end up paying for this boat after they split the rebate? (Remember, a **rebate** is money you get returned to you.)

a) ($5,600 + $1,200) ÷ 3
b) ($5,600 ÷ 3) − ($1,200 ÷ 3)
c) ($5,600 ÷ 3) + ($1,200 ÷ 3)
d) ($5,600 − $1,200) ÷ 3
e) ($1,200 ÷ 3) + $5,600

Overview of Computation Skills: Decimals, Fractions, and Percents

On the following four pages is an overview of the new computation skills you will study in Book 2. You may find that you are already familiar with many of these skills, while others may seem unfamiliar. This is perfectly natural.

The overview will help you evaluate your strengths and weaknesses in decimals, fractions, and percents. How you do on this overview may help you decide where you want to spend the most time in Chapters 2 through 6. However, if you're preparing to take a GED or other test, it is a good idea to work your way through the entire book, even reviewing those skills with which you are already familiar.

Take your time as you work the problems, and check your answers when you're finished. Answers appear in the answer key on page 186.

Decimal Skills: Study pages 34 through 65.

In each group of decimal fractions below, circle the one with the largest value.

1. .068, .580, .604   .70, .08, .39,   .0409, .0410, .049

Round each decimal fraction below to the nearest 10th place.

2. .66   .92   .07   .353   .7089

Add or *subtract* as indicated.

3.
.5	.63	.78	1.35	4.26	12.520
+.4	+.34	+.69	+ .67	+ 1.3	+ 8.25

4.
.9	.89	.385	3.18	5.722	53.006
−.6	−.7	−.286	− .89	−2.005	−27.059

Multiply or *divide* as indicated.

5.
$$\begin{array}{r} .45 \\ \times\ 6 \end{array} \qquad \begin{array}{r} .08 \\ \times\ 3 \end{array} \qquad \begin{array}{r} 4.7 \\ \times\ .8 \end{array} \qquad \begin{array}{r} .74 \\ \times .36 \end{array} \qquad \begin{array}{r} 4.502 \\ \times\ .26 \end{array} \qquad \begin{array}{r} 1,000 \\ \times\ 1.39 \end{array}$$

6. $6\overline{)2.4}$ $\qquad$ $14\overline{)32.76}$ $\qquad$ $5\overline{).265}$ $\qquad$ $.07\overline{)2.646}$ $\qquad$ $12.34 \div 100$

Fraction Skills: Study pages 66 through 113.

Change each improper fraction to a mixed number. Reduce every proper fraction.

7. $\dfrac{4}{6}$ $\qquad$ $\dfrac{21}{5}$ $\qquad$ $\dfrac{12}{9}$ $\qquad$ $\dfrac{6}{8}$ $\qquad$ $\dfrac{16}{6}$ $\qquad$ $\dfrac{10}{14}$

In each group of numbers below, circle the one with the largest value.

8. $\dfrac{3}{4}, \dfrac{7}{8}, \dfrac{1}{2}$ $\qquad$ $2\dfrac{4}{6}, \dfrac{15}{6}, 2\dfrac{15}{18}$ $\qquad$ $\dfrac{7}{5}, \dfrac{12}{10}, 1\dfrac{1}{5}$ $\qquad$ $\dfrac{3}{5}, \dfrac{3}{4}, \dfrac{3}{6}$

Express each amount below as a fraction of the larger unit indicated.

9. 8 inches = ____ foot $\qquad$ 25 minutes = ____ hour $\qquad$ 28 weeks = ____ year

Add or *subtract* as indicated.

10.
$$\begin{array}{r} \frac{5}{7} \\ +\frac{1}{7} \end{array} \qquad \begin{array}{r} \frac{4}{9} \\ +\frac{2}{9} \end{array} \qquad \begin{array}{r} 3\frac{1}{4} \\ +\ \frac{2}{4} \end{array} \qquad \begin{array}{r} 1\frac{7}{8} \\ +\ \frac{5}{8} \end{array} \qquad \begin{array}{r} 3\frac{5}{6} \\ +2\frac{1}{3} \end{array} \qquad \begin{array}{r} 13\frac{11}{12} \\ +\ 9\frac{3}{4} \end{array}$$

11.
$$\begin{array}{r} \frac{7}{8} \\ -\frac{2}{8} \end{array} \qquad \begin{array}{r} \frac{12}{15} \\ -\frac{8}{15} \end{array} \qquad \begin{array}{r} 6\frac{3}{4} \\ -\ \frac{1}{2} \end{array} \qquad \begin{array}{r} 9\frac{3}{8} \\ -5\frac{1}{4} \end{array} \qquad \begin{array}{r} 1\frac{1}{3} \\ -\ \frac{2}{3} \end{array} \qquad \begin{array}{r} 7\frac{1}{4} \\ -3\frac{5}{8} \end{array}$$

12.
$$\begin{array}{r} \frac{1}{3} \\ +\frac{1}{4} \end{array} \qquad \begin{array}{r} \frac{3}{5} \\ +\frac{3}{4} \end{array} \qquad \begin{array}{r} 4\frac{2}{3} \\ +2\frac{3}{4} \end{array} \qquad \begin{array}{r} \frac{7}{8} \\ -\frac{5}{12} \end{array} \qquad \begin{array}{r} 1\frac{2}{3} \\ -\ \frac{3}{4} \end{array} \qquad \begin{array}{r} 8\frac{2}{5} \\ -5\frac{2}{3} \end{array}$$

Multiply or *divide* as indicated.

13. $\frac{1}{2} \times \frac{3}{4}$ $\frac{6}{8} \times \frac{4}{5}$ $7 \times \frac{3}{4}$ $\frac{1}{3} \times 2\frac{3}{5}$ $3\frac{1}{2} \times 4\frac{3}{8}$

14. $\frac{5}{8} \div \frac{2}{7}$ $\frac{8}{9} \div \frac{3}{4}$ $\frac{4}{5} \div 6$ $2\frac{1}{2} \div \frac{2}{3}$ $1\frac{1}{3} \div 3\frac{5}{6}$

Percent Skills: Study pages 114 through 141.

Change each percent below to a decimal and a fraction. Write the decimal answer on the first line and the fraction answer on the second line.

15. 25% _____ _____   37.5% _____ _____ $33\frac{1}{3}$% _____ _____

Change each percent below to a whole number or a mixed decimal.

16. 100%   300%   450%   225%

Determine each number as indicated.

17. 50% of 90   90% of 38   25% of 64   2% of 14

18. .6% of 400   8.8% of 2,000 $\frac{3}{4}$% of 50 $33\frac{1}{3}$% of 57

Determine each percent as asked for below.

19. 12 is what percent of 60?   What percent of 64 is 8?

20. 250 is what percent of 50?   What percent of 30 is 10?

Determine each number as indicated.

21. 50% of what number is 47?   24 is 40% of what number?

22. .6 is 15% of what number? $33\frac{1}{3}$% of what number is 26?

Overview of Computation Skills Evaluation Chart

On the chart below, circle the number of the row in which you missed one or more answers. The skill and study pages associated with the problems in each row are indicated.

Row Number	Associated Skill	Study Pages
DECIMAL SKILLS		
1	Recognizing the value of decimal fractions	34 to 39
2	Rounding decimal fractions	40 to 41
3	Adding decimal fractions and mixed decimals	42 to 43
4	Subtracting decimal fractions and mixed decimals	44 to 45
5	Multiplying decimal fractions and mixed decimals	48 to 51
6	Dividing decimal fractions and mixed decimals	52 to 57
FRACTION SKILLS		
7	Recognizing the value of proper and improper fractions	66 to 67
8	Simplifying proper and improper fractions	68 to 71
9	Writing a part as a fraction of a whole	72
10	Adding fractions and mixed numbers	73 to 75
11	Subtracting fractions and mixed numbers	76 to 79
12	Finding common denominators for fraction addition and subtraction problems	82 to 87
13	Multiplying fractions and mixed numbers	92 to 95
14	Dividing fractions and mixed numbers	96 to 99
PERCENT SKILLS		
15	Changing percents to decimals and to fractions	114 to 117
16	Changing percents larger than 100% to whole numbers or to mixed numbers	118
17 & 18	Finding part of a whole	126 to 127
19 & 20	Finding what percent a part is of a whole	132
21 & 22	Finding a whole when part of it is given	136

2
Becoming Familiar with Numbers Smaller than 1

Using whole numbers, you know you can divide a group of objects into several smaller groups. For example, if you divide 30 pieces of candy among 6 children, each child gets 5 pieces. In mathematical symbols we write: $30 \div 6 = 5$.

You can also divide one whole object into smaller parts. In this case, each part is represented by a number smaller than 1.

As an example, the one large square at right is divided into 100 smaller squares. Each small square is **one hundredth** of the large square. The 27 shaded squares represent **27 hundredths** of the large square. Both 1 hundredth and 27 hundredths are numbers smaller than 1.

On the next four pages we'll briefly preview the 3 ways to mathematically write numbers that are smaller than 1:

- as **decimal fractions**
- as **common fractions**
- as **percents**

Decimal Fractions

Each of us sees or uses decimal fractions every day. This is because our money system uses them. Decimal fractions are used to represent parts of a dollar called *cents*. One cent (a penny) is one hundredth of a dollar. As a decimal fraction, hundredths are written as the first two numbers to the right of a decimal point.

Think of a dollar as being divided like the large square above. Twenty-seven hundredths of a dollar is 27¢. As a decimal fraction, 27¢ is written as follows:

27¢ = $.27   (Read each as "27 cents" although you can also correctly read
             $.27 as "27 hundredths of a dollar.")

└── decimal point

Two of the most common uses of decimal fractions are our money system and the metric measuring system. In the metric system, a meter is divided into 100 equal parts called **centimeters**. Because of this, a length such as 35 centimeters is often written as the decimal fraction .35 meter (35 hundredths of a meter).

EXAMPLE USES OF DECIMAL FRACTIONS

$.25 $.10 $.05 $.01   .35 meter

Common Fractions

A common fraction also represents part of a whole. A common fraction is written as one number above a second number. For example, 27¢ is part of one dollar and can be written as a common fraction:

$$27¢ = \frac{27}{100} \text{ of a dollar} \quad \text{(Read as "27 hundredths of a dollar.")}$$

A common fraction has two parts. The top number is called the **numerator**, and the bottom number is called the **denominator**.

$\frac{27}{100}$ ← numerator   The **numerator** tells how many parts you have.

← denominator   The **denominator** tells how many parts the whole is divided into.

Common fractions are used in check writing, in measurement, and in cooking recipes. As an example, a 12-inch ruler is divided into inches, each of which is divided into $\frac{1}{16}$-, $\frac{1}{8}$-, $\frac{1}{4}$-, and $\frac{1}{2}$-inch units.

EXAMPLE USES OF COMMON FRACTIONS

$\frac{13}{16}$"

RECIPE

$\frac{1}{2}$ cup of honey

$\frac{1}{4}$ cup of oil

$\frac{3}{8}$ cup of currants

Percents

Percent means "hundredth." A percent is most often written as a number followed by the percent symbol %. For example, 5 percent is usually written 5% and means 5 hundredths.

In our money system, 1 penny is 1% of a dollar, since 1 penny is 1 hundredth of a dollar. In the same way, 27¢ is 27% of a dollar.

27¢ = 27% of a dollar   (Read as "27 percent of a dollar.")
                ⌐————— % means hundredth

The most common uses of percents are seen in newspaper and magazine advertisements. Sale prices are usually given in percents. Money interest rates, both the rates paid to savers and the rates charged to borrowers, are given as percents.

EXAMPLE USES OF PERCENTS

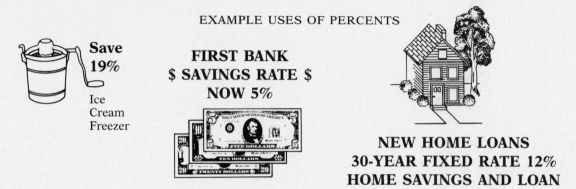

Save 19%

Ice Cream Freezer

FIRST BANK
$ SAVINGS RATE $
NOW 5%

NEW HOME LOANS
30-YEAR FIXED RATE 12%
HOME SAVINGS AND LOAN

As you've seen, a number smaller than 1 can be written in any one of three ways. In the examples below, we show how decimal fractions, common fractions, and percents are used to write cents as part of a dollar.

Examples:

Cents	Written as a decimal fraction	Written as a common fraction	Written as a percent
1¢	$.01	$\frac{1}{100}$ of a dollar	1% of a dollar
27¢	$.27	$\frac{27}{100}$ of a dollar	27% of a dollar
90¢	$.90	$\frac{90}{100}$ of a dollar	90% of a dollar

Complete the following chart to show the three equally correct ways of writing cents as part of a dollar. Several blanks are filled in for you.

	Cents	Written as a decimal fraction	Written as a common fraction	Written as a percent
1.	3¢			3% of a dollar
2.	8¢			
3.	14¢	$.14		
4.	29¢			
5.	40¢			
6.	72¢		$\frac{72}{100}$ of a dollar	
7.	10¢			

8.	85¢			
9.	99¢			
10.	6¢			

On the line above each picture, write *decimal fraction*, *common fraction*, or *percent* to identify which is being used in that picture.

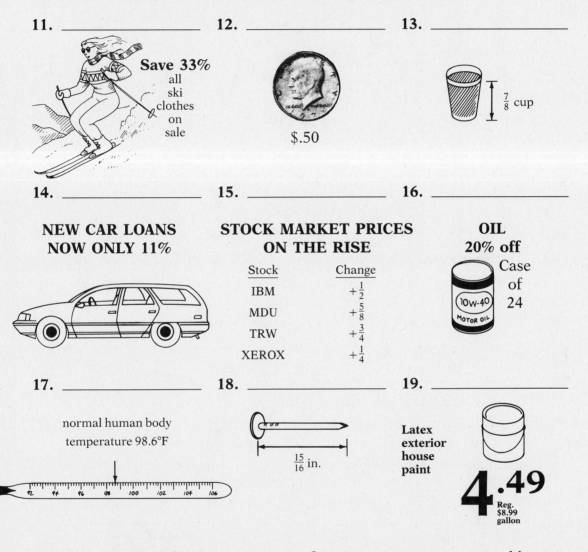

11. _____

Save 33%
all
ski
clothes
on
sale

12. _____

$.50

13. _____

7/8 cup

14. _____

**NEW CAR LOANS
NOW ONLY 11%**

15. _____

**STOCK MARKET PRICES
ON THE RISE**

Stock	Change
IBM	+$\frac{1}{2}$
MDU	+$\frac{5}{8}$
TRW	+$\frac{3}{4}$
XEROX	+$\frac{1}{4}$

16. _____

**OIL
20% off**

10W-40
MOTOR OIL

Case
of
24

17. _____

normal human body
temperature 98.6°F

92 94 96 98 100 102 104 106

18. _____

$\frac{15}{16}$ in.

19. _____

**Latex
exterior
house
paint**

4.49
Reg.
$8.99
gallon

Which — **decimal fraction**, **common fraction**, or **percent** — would you expect to find each of the following people working with the most?

20. A baker _____

21. A sales clerk _____

22. A car salesman _____

23. A cashier _____

24. A stockbroker _____

25. A carpenter _____

26. A school nurse _____

27. A real estate
loan agent _____

33

3
Decimal Skills

Introducing Mixed Decimals

In the previous chapter, we pointed out that our money system uses **decimal fractions**.

- Cents are hundredths of a dollar. Cents are written as the first two numbers to the right of the decimal point.

- Dollars are written as whole numbers to the left of the decimal point.

$7.29

└──── The decimal point separates whole numbers from the decimal fraction.

The number 7.29 is an example of a **mixed decimal**. A mixed decimal is a whole number plus a decimal fraction:

mixed decimal = whole number + decimal fraction

When you read a mixed decimal, you read the decimal point as the word "and." This is true both for money and for a number standing alone:

$7.29 is read as "Seven dollars *and* twenty-nine cents."

7.29 is read as "Seven *and* twenty-nine hundredths."

Each place in a mixed decimal has a certain **place value**. To the left of the decimal point are the familiar place values of whole numbers.

To the right of the decimal point are the place values of decimal numbers. The first place is the **tenths place**. To the right of the tenths place is the **hundredths place**. Continuing to the right, each new place value decreases by a multiple of ten. Notice that each decimal place value ends in the letters *ths*.

You are already familiar with tenths and hundredths places from your use of money:

- The number of dimes is written in the tenths place. A dime is one tenth of a dollar.

- The number of pennies is written in the hundredths place. A penny is one hundredth of a dollar, but it is only one tenth of a dime.

As you study **decimals** (decimal fractions and mixed decimals) you'll want to become familiar with the first six whole number and decimal fraction place value names.

PLACE VALUE NAMES IN MIXED DECIMALS

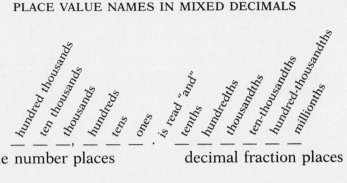

whole number places   decimal fraction places

It is helpful to remember that each decimal place stands for part of a whole. Look at the meaning of each of these simple decimal fractions:

Example	Value	Meaning
.1	one tenth	one part out of 10 parts
.01	one hundredth	one part out of 100 parts
.001	one thousandth	one part out of 1,000 parts
.0001	one ten-thousandth	one part out of 10,000 parts
.00001	one hundred-thousandth	one part out of 100,000 parts
.000001	one millionth	one part out of 1,000,000 parts

The number of digits to the right of the decimal point is called the **number of decimal places**. Here's a good way to remember decimal place names:

- 10 has one zero, and tenth has one decimal place.

- 100 has two zeros, and hundredth has two decimal places.

- 1,000 has three zeros, and thousandths has three decimal places, and so on.

Use words to express the value of each **whole number** and **decimal fraction** below.

1. 1,000 _____ .001 _____

2. 100 _____ .01 _____

3. 10 _____ .1 _____

4. 100,000 _____ .00001 _____

5. 10,000 _____ .0001 _____

Express the following **mixed decimals** in words.

6. 2.1 _____ **7.** 4.1 _____

8. 5.01 _____ **9.** 7.001 _____

10. $2.10 _____ **11.** $13.01 _____

35

Reading Decimals

A decimal fraction may have one or more nonzero digits. To read a decimal fraction, first read the number to the right of the decimal point. Read this number just as you'd read a whole number. Then read the place value of the digit farthest to the right. **You can think of a decimal fraction as a number plus a place value**.

Examples	Number +	Place Value	Read as
.4	4	tenths	4 tenths
.04	4	hundredths	4 hundredths
.35	35	hundredths	35 hundredths
.035	35	thousandths	35 thousandths
.140	140	thousandths	140 thousandths

Although the number 4 in .4 is the same as the number 4 in .04, these decimal fractions have different values. The 4 in .4 is in the tenths place, while the 4 in .04 is in the hundredths place.

Similarly, .35 differs from .035. Each has number 35. But the place value in each is determined by the 5, the farthest digit to the right. In .35 the place value is hundredths. In .035 the place value is thousandths.

When reading a mixed decimal, remember to read the decimal point as the word *and*, connecting the whole number with the decimal fraction.

17.538 is read "17 *and* 538 thousandths."

Show how each of the following decimal fractions is read. Two are done for you.

1. .5 _5 tenths_   .7 _____   .9 _____

2. .12 _12 hundredths_ .27 _____   .50 _____

3. .135 _____   .272 _____   .180 _____

4. .2048 _____   .1305 _____   .4000 _____

5. .12348 _____   .83721 _____   .204000 _____

Show how each of the following mixed decimals is read.

6.   5.8 _____   7.19 _____

7.   6.87 _____   14.105 _____

8. 29.326 _____   154.60 _____

9.   3.2442 _____   18.38724 _____

Writing Zero as a Place Holder

Although the digit 0 has no value, it is used as a **place holder**.

- Placed between the decimal point and a decimal digit, zero changes the value of a decimal fraction.

- Placed at the far right of a decimal fraction, zero changes the way the fraction is read but does not change its value.

Example 1: The decimal fraction .04 differs from .4 because of the 0 in the tenths place. It is this 0 that *holds* 4 in the hundredths place.

> A zero that comes anywhere between the decimal point and the last nonzero digit is called a **necessary zero**. A necessary zero cannot be removed without changing the value of a number.
>
> **Examples of necessary zeros:** .0̱4   .5̱06   3.0̱205   .0̱080

Example 2: The decimal fractions .60 and .6 differ in the way they are read. You read .60 as 60 hundredths, and .6 as 6 tenths. Yet both have the same value. This is similar to the fact that 60 pennies have the same value as 6 dimes.

> Because zeros at the far right do not change the value of a decimal fraction, these zeros are called **unnecessary zeros**.
>
> **Examples of unnecessary zeros:** .6̱0   .25̱00   4.703̱0

Underline each **necessary zero** in the decimals below. Circle each **unnecessary zero**.

1. .05	.106	.007	2.30	4.650
2. .109	.048	5.070	.2031	5.0040
3. 2.0030	.650	.07700	1.1030	.50050

In each group of three decimal fractions below, circle the two that have the same value.

4. .03, .003, .030	.405, .450, .45	.61, .061, .610
5. .015, .0150, .105	.0271, .2710, .271	.0306, .03060, .30600

Writing Decimals

To write a decimal fraction, first decide the **place value** of your number. The place value tells you how far from the decimal point to place the number's last digit.

Example: Write **thirty-six thousandths** as a decimal fraction.

Step 1. Identify **thousandths** as the place value. You know this because thousandths is the last word of the number. Also, place value is always a word that ends in *th* or *ths*.

Step 2. Write 36. Place the decimal point so that 6 ends up in the thousandths place, the third place to the right. Since 36 is only a two-digit number, use a zero to hold the first place.

┌─0 is written as a place holder.
.036
└─6 ends up in the thousandths place.

Answer: thirty-six thousandths is written as .036

To write a mixed decimal, remember that the word *and* separates the whole number from the decimal fraction.

Examples	*Written as a decimal*
five hundredths	.05
twenty-one hundredths	.21
sixty-four thousandths	.064
four and three hundred seventeen thousandths	4.317
five and seven tenths	5.7
ninety and sixty-one hundredths	90.61

Write each number below as a decimal fraction or as a mixed decimal.

1. nine tenths _____

2. seven hundredths _____

3. thirty-five hundredths _____

4. twenty-seven thousandths _____

5. four and one tenth _____

6. two and nine hundredths _____

7. fifty and twenty hundredths _____

8. six and eighty thousandths _____

9. eight hundred forty-three thousandths _____

10. one hundred seventy-four and sixty-five hundredths _____

11. one thousand four hundred and seventy-five thousandths _____

Comparing Decimal Fractions

Decimal fractions can easily be compared when they have the same number of decimal places. For example, you know that $.40 is larger than $.08. Since each amount is written with two decimal places, you simply compare the number 40 with the number 8.

This idea can be used to compare any decimal fractions.

Rules for Comparing Decimal Fractions

1. Use zeros to give each decimal fraction the same number of places. Remember, placing one or more zeros **to the right** of a decimal fraction does not change its value.
2. Compare the numbers.

Example: Which decimal fraction is larger, .07 or .048?

Step 1. Give .07 and .048 the same number of places. To do this, place a zero at the right end of .07: .07 = .070.

Step 2. Compare the numbers 70 and 48. Because 70 is larger than 48, .070 is larger than .048. Therefore, .07 is larger than .048.

Answer: **.07 is larger than .048**

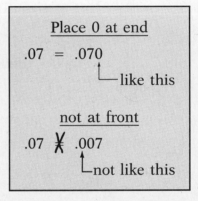

In each pair below, circle the larger decimal fraction.

1. .32 or .51	**2.** .72 or .39	**3.** .137 or .401
4. .29 or .4	**5.** .3 or .184	**6.** .66 or .493
7. .103 or .42	**8.** .76 or .2321	**9.** $.23 or $.09
10. .482 or .2673	**11.** .392 or .6004	**12.** $.80 or $.56

Arrange each group of numbers below in order. In each group write the smallest number to the left and the largest to the right. The first one is done for you.

13. .43, .8, .134
.430, .800, .134
.134, .43, .8

14. .201, .4, .35

15. $.53, $.09, $.28

16. .209, .45, .5

17. .6, .35, .42

18. .611, .64, .174

Rounding Decimal Fractions

To **round a decimal fraction** is to simplify the way it is written. You do this by discarding digits that are not needed.

Example 1: Erin earns $7.68 for each hour of overtime she works. How much will she earn in 2.4 hours of overtime work on Saturday?

Step 1. You solve this problem by multiplying $7.68 by 2.4. This multiplication is shown at right.

You'll learn how to multiply decimals later in this chapter. For now, notice that the answer contains 3 digits to the right of the decimal point.

$$\begin{array}{r} \$7.68 \\ \times\ \ \ 2.4 \\ \hline 3\ 072 \\ 15\ 36 \\ \hline \$18.432 \end{array}$$

Step 2. To write $18.432 as dollars and cents, we want to keep only two digits to the right of the decimal point. To do this, we **round** $18.432 to $18.43 by discarding the 2.

Answer: $18.43

The steps for rounding decimal fractions are almost the same as the steps for rounding whole numbers.

Steps for Rounding Decimal Fractions

1. Underline the digit in the place you are rounding to.

2. Look at the digit to the right of the underlined digit:

   a) If the digit to the right is 5 or more, add 1 to the underlined digit.

   b) If the digit to the right is less than 5, leave the underlined digit as it is.

3. Discard the digits to the right of the underlined digit.

Example 2: Round 2.1749 to the thousandths place.

Step 1. Underline the digit in the thousandths place.
Underline the 4: 2.17_49.

Step 2. Look at the digit to the right of the 4. The digit is 9. Since 9 is "5 or more," add 1 to the underlined digit 4.
Add 1 to 4: 2.175̷9.

Step 3. Discard the digit 9.

Answer: 2.175

Round each amount below to the nearest cent. For each amount circle one of the two answer choices.

1. $.467: $.46 *or* $.47 $.015: $.01 *or* $.02 $.953: $.95 *or* $.96

2. $6.875: $6.87 *or* $6.88 $9.023: $9.02 *or* $9.03

3. $23.252: $23.25 *or* $23.26 $83.746: $83.74 *or* $83.75

Round each decimal fraction below as indicated. The first problem in each row is done for you.

To the nearest 10th

4. .24 _.2_   .37 _____   .52 _____   .408 _____   .375 _____

.24
⤷ less
than 5

To the nearest 100th

5. .406 _.41_   .483 _____   .725 _____   .8715 _____   .3842 _____

.40̲6
⤷ 5 or
more

To the nearest 1000th

6. .0273 _.027_   .8394 _____   .3847 _____   .6283 _____   .93042 _____

.027̲3
⤷ less
than 5

7. One inch equals 2.54 centimeters. Round this length to the nearest tenth of a centimeter.

8. A ⅜-inch drill bit has a diameter of .375 inches. What size is this to the nearest hundredth inch?

9. When regular gas is selling for $1.069 per gallon, what is its price to the nearest cent per gallon?

10. The machinist's blueprint showed that the key should be 2.136 inches long. What length is this to the nearest hundredth inch?

Adding Decimals

To add decimals, line up the decimal points and then add each column just as you do when you add whole numbers. **Place a decimal point in the answer directly below the decimal points in the problem.**

Example 1: Add .23 and .16.

    Step 1. Line up the decimal points and add the columns.

$$\begin{array}{r} .23 \\ +.16 \\ \hline .39 \end{array}$$

    Step 2. Place the decimal point in the answer directly below the decimal points.

Answer: **.39**

Example 2: Add 5.23 and 2.14.

$$\begin{array}{r} 5.23 \\ +2.14 \\ \hline 7.37 \end{array}$$

Answer: **7.37**

Carrying with decimals is done the same way as carrying with whole numbers. **Carry across the decimal point as if it weren't there.**

Example 3: Add .85 and .63.

    Step 1. Line up the decimal points and add the columns. Carry the 1 from the tenths column to the ones column.

$$\begin{array}{r} {}^{1}.85 \\ +.63 \\ \hline 1.48 \end{array}$$

    Step 2. Place the decimal point in the answer.

Answer: **1.48**

Example 4: Add 9.64 and 4.58.

$$\begin{array}{r} {}^{1\,1\,1}9.64 \\ +4.58 \\ \hline 14.22 \end{array}$$

Answer: **14.22**

Add.

1.
$$\begin{array}{r} .3 \\ +.2 \\ \hline \end{array} \qquad \begin{array}{r} .42 \\ +.16 \\ \hline \end{array} \qquad \begin{array}{r} .624 \\ +.304 \\ \hline \end{array} \qquad \begin{array}{r} 2.1 \\ +1.7 \\ \hline \end{array} \qquad \begin{array}{r} 23.4 \\ +12.3 \\ \hline \end{array} \qquad \begin{array}{r} 5.376 \\ +3.210 \\ \hline \end{array}$$

2.
$$\begin{array}{r} .9 \\ +.7 \\ \hline \end{array} \qquad \begin{array}{r} .84 \\ +.32 \\ \hline \end{array} \qquad \begin{array}{r} .645 \\ +.571 \\ \hline \end{array} \qquad \begin{array}{r} 6.8 \\ +5.5 \\ \hline \end{array} \qquad \begin{array}{r} 8.94 \\ +4.46 \\ \hline \end{array} \qquad \begin{array}{r} 23.85 \\ +14.98 \\ \hline \end{array}$$

When adding money, write $ next to only the top number and the answer.

3.
$$\begin{array}{r} \$.34 \\ +.25 \\ \hline \end{array} \qquad \begin{array}{r} \$.76 \\ +.32 \\ \hline \end{array} \qquad \begin{array}{r} \$.89 \\ +.46 \\ \hline \end{array} \qquad \begin{array}{r} \$1.63 \\ +1.50 \\ \hline \end{array} \qquad \begin{array}{r} \$21.96 \\ +14.37 \\ \hline \end{array} \qquad \begin{array}{r} \$39.57 \\ +26.85 \\ \hline \end{array}$$

Zeros as Place Holders

To add numbers that do not have the same number of decimal places, you may find it helpful to use zeros as place holders. Extra zeros keep columns in line as you add. A whole number is "understood" to have a decimal point to its immediate right.

Example 5: Add 8, 5.231, and 1.36.

		Lining up the decimal points	Using zeros as place holders
Step 1.	Write a decimal point to the right of the 8, and line up the three decimal points. Use zeros to give each number the same number of decimal places.	8. 5.231 + 1.36	8.000 5.231 + 1.360 14.591
Step 2.	Add the columns.		

Answer: 14.591

To use zeros as place holders, you must write them at the end of the decimal fractions. In this way you do not change the values of the numbers being added.

Use zeros as place holders to add each problem below.

1.
$$.56 \\ +.2$$
$$.6 \\ +.14$$
$$.232 \\ +.16$$
$$.743 \\ +.2$$
$$.864 \\ +.76$$
$$.92 \\ +.7507$$

2.
$$1.03 \\ +.8$$
$$2. \\ +1.3$$
$$3.21 \\ +1.6$$
$$5.6 \\ +3.$$
$$2.735 \\ +1.55$$
$$12.85 \\ +6.1$$

Perform each addition below. Remember to place a decimal point to the right of each whole number. Use extra zeros as place holders to keep columns in line.

3. .34 + .12   .64 + .5   .345 + .25   .2535 + .12

4. 4 + 3.26   2.175 + .92   5.21 + 3   6.81 + 4.2

5. 7 + 3.61 + 1.3   4.2 + 3 + 2.87   6.81 + 4 + 3.9

Subtracting Decimals

To subtract decimals, place the smaller number directly below the larger number and line up the decimal points. Then subtract, borrowing if necessary, just as you would subtract whole numbers. **Place a decimal point in the answer directly below the decimal points in the problem.**

Example 1: Subtract 1.24 from 5.61.

Step 1. Write 1.24 directly below 5.61. Make sure you line up the decimal points. Subtract by borrowing from the tenths column in order to subtract the hundredths column.

$$\begin{array}{r} 5.\overset{5\ 11}{\cancel{6}\cancel{1}} \\ -1.24 \\ \hline 4.37 \end{array}$$

Step 2. Place a decimal point in the answer.

Answer: 4.37

Subtract. The first problem in each row is worked as an example.

1.
.7	.9	.8	.75	.97	2.58
−.4	−.3	−.1	−.62	−.25	−.26
.3					

2.
2.6	7.4	9.45	13.9	4.76	23.67
−1.3	−3.2	−3.14	−5.6	−2.53	−7.50
1.3					

3.
2.$\overset{2\ 15}{\cancel{3}\cancel{5}}$	6.76	$8.07	14.554	7.10	25.03
−1.17	−4.49	−4.43	−7.982	−5.78	−9.57
1.18					

Rewrite each problem below by placing the smaller number directly below the larger number. Line up the decimal points and subtract.

4. .9 − .7   .8 − .4   1.83 − .92   12.34 − 9.28

$$\begin{array}{r} .9 \\ -.7 \\ \hline .2 \end{array}$$

5. $3.75 − $2.68 $12.29 − $7.86 $7.85 − $4.59   2.456 − 1.488

$$\begin{array}{r} \overset{6\ 15}{\cancel{3}.\cancel{75}} \\ -2.68 \\ \hline \$1.07 \end{array}$$

To subtract numbers that do not have the same number of decimal places, write in extra zeros as place holders. When necessary, borrow from these zeros in the same way you borrow when subtracting whole numbers. You can borrow across the decimal point as if it weren't there.

Example 3: Subtract 3.4 from 5.82.   **Example 4:** Subtract 2.36 from 9

Lining up the   *Using a zero as*   *Lining up the*   *Using two zeros*
decimal points   *a place holder*   *decimal points*   *as place holders*

$$\begin{array}{r} 5.82 \\ -3.4 \\ \hline \end{array} \qquad \begin{array}{r} 5.82 \\ -3.40 \\ \hline 2.42 \end{array} \qquad \begin{array}{r} 9. \\ -2.36 \\ \hline \end{array} \qquad \begin{array}{r} \overset{8\ 9\ 10}{9.\cancel{0}\cancel{0}} \\ -2.36 \\ \hline 6.64 \end{array}$$

Answer: 2.42   **Answer: 6.64**

Use zeros as place holders to subtract in each problem below.

6.
$$\begin{array}{r} .89 \\ -.5 \\ \hline \end{array} \quad \begin{array}{r} .74 \\ -.5 \\ \hline \end{array} \quad \begin{array}{r} .972 \\ -.5 \\ \hline \end{array} \quad \begin{array}{r} 1.45 \\ -1.3 \\ \hline \end{array} \quad \begin{array}{r} 3.45 \\ -1.8 \\ \hline \end{array} \quad \begin{array}{r} 12.703 \\ -8.4 \\ \hline \end{array}$$

7.
$$\begin{array}{r} 2.76 \\ -1.8 \\ \hline \end{array} \quad \begin{array}{r} 3.67 \\ -1.9 \\ \hline \end{array} \quad \begin{array}{r} 5.75 \\ -2.8 \\ \hline \end{array} \quad \begin{array}{r} .467 \\ -.29 \\ \hline \end{array} \quad \begin{array}{r} 3.40 \\ -1.83 \\ \hline \end{array} \quad \begin{array}{r} 42.05 \\ -17.37 \\ \hline \end{array}$$

Perform each subtraction below. Place a decimal point to the right of each whole number and write in extra zeros as needed. Be sure to line up the decimal points.

8. .95 − .8   .83 − .7   1.45 − .3   3.58 − 1.2

9. 1.78 − .9   2.56 − 1.8   .3802 − .27   .504 − .3

10. 7 − 5.37   4 − 3.48   3.53 − 2   12 − 4.598

11. .304 − .10   3.46 − 3.1   8 − 4.204

Solving Addition and Subtraction Word Problems

In each of problems 1, 2, and 3, underline a key word that suggests either addition or subtraction. Then solve each problem.

1. During the first half of baseball season, Hank's batting average was .306. During the second half, his average increased by .048. What was Hank's batting average at the end of the season?

2. The area of the United States is 3.62 million square miles, while the area of the Soviet Union is 8.57 million square miles. What is the difference in area of these two countries?

3. For lunch, Lucy had a ham sandwich for $2.48, a piece of apple pie for $1.29, and a cup of coffee for $.75. If the sales tax came to $.23, what is Lucy's total lunch bill?

Problems 4, 5, and 6 contain extra information. In each problem, circle only the necessary information and then solve the problem.

4. The distance from Joe's house to his workplace is 12.8 miles. His workplace is 5.6 miles north of town. On the way to work he drops his daughter off at school, a distance of 5.9 miles from his house. After he drops her off, how much farther does Joe have to drive to get to his workplace?

5. Georgia bought a loaf of bread for $.79. She paid with a five-dollar bill and was given $3.81 in change. Seeing that he'd made a mistake, the clerk gave Georgia more money. What should Georgia's correct change be?

6. A machinist cut .025 inch off a shaft that first measured .752 inch across. Since it was still a little too wide, he cut another .003 inch off. Now the shaft fits perfectly. How much did the machinist end up cutting off the shaft?

In problems 7 and 8, circle the arithmetic expression that will give the correct answer to each question. You do not need to solve these two problems.

7. Normal human body temperature is 98.6 degrees. When she had the flu, Jill's temperature went up to 103.4 degrees. By taking a cool bath, Jill lowered her temperature to 101.7. How much did the bath reduce Jill's temperature?

a) 103.4 − 98.6
b) 103.4 − 101.7
c) 101.7 − 98.6
d) 103.4 + 101.7

8. The new Japanese Aikeo sports car measures 5.42 meters long. A similar American car measures 5.63 meters. Aikeo's new model is .135 meters shorter than last year's Aikeo sport model. Knowing this, determine the length of last year's Aikeo sports car.

a) 5.42 − .135
b) 5.63 + .135
c) 5.63 + .135
d) 5.42 + .135

In questions 9, 10, and 11, complete the drawings by placing numbers from the problems on the blank lines. Then use the drawings to help you solve each problem.

9. Bolt A is .465 inch longer than bolt B. Bolt B is .289 inch longer than bolt C. How much longer is bolt A than bolt C?

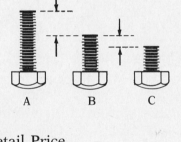

A   B   C

10. The retail price of a wool sweater at Jessica's Clothes Shop is $59.75. As the owner, Jessica can buy the sweater herself anytime for a price of $39.99. How much markup does Jessica have on this sweater?

Retail Price   _____
  subtract
Owner's Price   _____
Equals Markup   _____

11. Sid drew the map at right and measured the following dimensions for his property:

north side: 53.75 meters
east side: 79.69 meters
south side: 51.62 meters
west side: 84.39 meters

Walking clockwise around the eastern edge of the property, how far is it from the northwest corner A to the southwest corner B?

____N

A

W   ____E

B

____S

Multiplying Decimal Numbers

When decimal numbers are multiplied, the number of decimal places in the answer must equal the sum of decimal places in the problem.

Example: Multiply .385 by .7.

Step 1. Multiply the numbers.

Step 2. Count the number of decimal places in each number being multiplied.

$$\begin{array}{r} .385 \\ \times \quad .7 \\ \hline 2695 \end{array} \; + \; \begin{array}{l} \text{three places} \\ \text{one place} \\ \hline \text{four places} \end{array}$$

Step 3. Add these two numbers (3 + 1 = 4) to see how many decimal places belong in the answer.

Starting at the right, count over 4 places to the left and place a decimal point in the answer.

.2695
4 3 2 1

Start at the right.
Count to the left.

Answer: **.2695**

Remember, the number of decimal places in a number is simply the number of digits to the right of the decimal point. A whole number has no decimal places.

Find each product below. First complete each row of partially worked Skill Builders **by placing a decimal point in each answer.**

Multiplying Decimals by Whole Numbers

Skill Builders

1.
| $\begin{array}{r}.35 \\ \times \quad 4 \\ \hline 140\end{array}$ two places / no places / two places | $\begin{array}{r}87 \\ \times .06 \\ \hline 522\end{array}$ no places / two places / two places | $\begin{array}{r}25 \\ \times .9 \\ \hline 225\end{array}$ no places / one place / one place | $\begin{array}{r}390 \\ \times .005 \\ \hline 1950\end{array}$ no places / three places / three places |

2.
| $\begin{array}{r}.67 \\ \times \quad 3\end{array}$ | $\begin{array}{r}.82 \\ \times \quad 6\end{array}$ | $\begin{array}{r}.09 \\ \times \quad 9\end{array}$ | $\begin{array}{r}1.5 \\ \times \quad 4\end{array}$ | $\begin{array}{r}2.8 \\ \times \quad 5\end{array}$ |

3.
| $\begin{array}{r}8.5 \\ \times \quad 6\end{array}$ | $\begin{array}{r}.74 \\ \times \quad 7\end{array}$ | $\begin{array}{r}126 \\ \times \quad .4\end{array}$ | $\begin{array}{r}23 \\ \times .08\end{array}$ | $\begin{array}{r}.89 \\ \times \quad 6\end{array}$ |

4.
$.35	215	$1.25	137	$5.17
× 8	×.004	× 5	×.008	× 6

Multiplying Decimals by Decimals

5.

.89 two places	2.3 one place	2.84 two places	.759 three places
× .7 one place	× .8 one place	× .76 two places	× .83 two places
623	184	1704	2277
		1988	6072
		21584	62997

6.
.46	.32	.75	1.5	3.7
× .8	× .6	× .9	× .9	× .5

7.
6.9	8.21	.77	2.83	.515
× .3	× .07	× .8	× .09	× .6

8.
1.26	.582	.903	4.74	3.89
× .32	× 6.6	× .72	× 1.6	× .35

Rounding Decimal Numbers

Round each answer to the place indicated. Turn back to page 40 to review how to round decimal numbers. The first problems are done for you.

To the nearest tenth.

9.
.24	.46	1.6	.325	4.61
× 8	× 4	× .9	× .8	× 6

1.92

= 1.92 ≈ 1.9

To the nearest hundredth.

10.
4.16	5.28	.35	.52	7.24
× .8	× .6	× .7	× .9	× .06

3.328

= 3.328 ≈ 3.33

In the following example, it is necessary to write a zero as a place holder before the decimal point can be placed in the answer. Can you see why?

Example: What is the product of .23 times .14?

$$
\begin{array}{r}
.23 \text{ two places} \\
\times .14 \text{ two places} \\
\hline
92 \\
23 \\
\hline
.0322 \text{ four places}
\end{array}
$$

partial products

Multiplying .23 by .14 gives a 3-digit number, but the answer must have 4 decimal places. This is why we add a zero to the left of the 3 before placing the decimal point in the answer.

It is not necessary to place decimal points in the partial products. It is only necessary to place one in the final answer.

Compute each product below. As a first step, complete the Skill Builders by correctly placing a decimal point in each answer after writing needed zeros.

Skill Builders

11.

.4 one place	.03 two places	.24 two places	.0009 four places
× .2 one place	× .8 one place	× .03 two places	× 7 no places
8 two places	24 three places	72 four places	63 four places

12.

.8	.7	.9	.57	.63
× .2	× .4	× .8	× .8	× .02

13.

.07	.05	.008	.015	.106
× .7	× .5	× .07	× .003	× .007

14.

.0007	.0006	7.2	.003	6.5
× 6	× 3	× .004	× .08	× .009

Multiplying by 10, 100, or 1,000

Decimal multiplication gives us three shortcuts to use when multiplying by 10, 100 or 1,000.

1. **To multiply a decimal by 10, move the decimal point one place to the right.**

   Example 1: Multiply 2.41 by 10.
   2.41 × 10 = 2.41
   = **24.1**

   Example 2: Multiply .09 by 10.
   .09 × 10 = .09
   = **.9**

2. **To multiply a decimal by 100, move the decimal point two places to the right.**

   Example 3: Multiply 5.6 times 100.
   5.6 × 100 = 5.60
   = **560**

   Example 4: Multiply:
   .3 × 100
   .3 × 100 = .30
   = **30**

Notice in examples 3 and 4 that zeros have to be added to the right of each number. These zeros hold places so that we can move the decimal point the correct number of places. As you see, these added zeros become part of the whole number answer.

3. **To multiply a decimal by 1,000, move the decimal point three places to the right.**

   Example 5: Multiply 13.7 by 1,000.
   13.7 × 1,000 = 13.700 = **13,700**

Hint: To remember these rules, notice that the decimal point is always moved the same number of places as there are 0s in the multiplying number.

Using the shortcuts, compute each product below. The first one in each row is done as an example.

1.
1.34	68.3	2.74	32.1	.058
× 10	× 10	× 10	× 10	× 10
13.4				

2. .35 × 10 = **3.5** .35   2.3 × 10 =   .03 × 10 =

3. .253 × 100 = **25.3** .253   .03 × 100 =   21.7 × 100 =

4. 5.14 × 1,000 = **5,140** 5.140   .05 × 1,000 =   74.9 × 1,000 =

Dividing a Decimal by a Whole Number

To divide a decimal by a whole number, place a decimal point in the answer **directly above its position in the problem**. Then divide the numbers just as you would divide whole numbers.

Example: Divide 3.76 by 4.

Step 1. Set up the problem for long division. Place a decimal point in the answer space above the line.

Step 2. Divide as you would divide whole numbers.

Answer: .94

$$\begin{array}{r} .94 \\ 4\overline{)3.76} \\ \underline{3\ 6} \\ 16 \\ \underline{16} \\ 0 \end{array}$$

Solve each division problem below. Start by completing the row of partially worked Skill Builders.

Skill Builders

1. $4\overline{)8.48}$ $5\overline{)45.5}$ $6\overline{)4.32}$ $8\overline{)7.408}$ $12\overline{)25.2}$
---

2. $3\overline{)6.3}$ $7\overline{)4.97}$ $2\overline{).842}$ $5\overline{)3.05}$ $6\overline{).612}$

3. $5\overline{)4.65}$ $9\overline{)20.7}$ $6\overline{)1.452}$ $14\overline{)32.76}$ $20\overline{)132.0}$

When dividing money, be sure to place a dollar sign in the answer.

4. $4\overline{)\$43.36}$ $14\overline{)2.884}$ $16\overline{)\$139.52}$ $6\overline{).936}$ $8\overline{)\$9.44}$

Using Zeros When You Can't Divide

As shown at right, a zero is used to hold a place when you can't divide:

Example:

$$4\overline{).236} \qquad \begin{array}{r} .059 \\ 4\overline{).236} \\ \underline{20} \\ 36 \\ \underline{36} \end{array}$$

Since you can't divide 4 into 2, put a 0 above the 2. Now divide 4 into 23. Place 5 above the 3 and continue the steps of long division.

Skill Builders

5.

$$7\overline{)\overset{.0}{.427}} \qquad 5\overline{)\overset{.0}{.305}} \qquad 4\overline{)\overset{.0}{.056}} \qquad 8\overline{)\overset{.0}{.0928}} \qquad 6\overline{)\overset{.00}{.0174}}$$

---

6. $8\overline{).648}$ $3\overline{).096}$ $7\overline{).0084}$ $12\overline{).108}$ $25\overline{).075}$

7. $9\overline{).297}$ $6\overline{).0774}$ $13\overline{).1053}$ $27\overline{).0567}$ $4\overline{).0504}$

Zeros can also be added to the end of a number to make division possible. At right, a zero is added to 2.4 to give 2.40. Now we can divide by 60.

Example:

$$60\overline{)2.4} \rightarrow \begin{array}{r} .04 \\ 60\overline{)2.40} \\ \underline{2\ 40} \end{array}$$

Skill Builders

8.

$$4\overline{).2} \rightarrow 4\overline{)\overset{.}{.20}} \qquad 20\overline{).16} \rightarrow 20\overline{)\overset{.}{.160}} \qquad 24\overline{).012} \rightarrow 24\overline{)\overset{.}{.0120}}$$

---

9. $6\overline{).3}$ $8\overline{).4}$ $32\overline{)1.6}$ $40\overline{)2.4}$ $60\overline{)1.2}$

Adding zeros is also useful in many other division problems. As shown in the first example at right, using zeros makes it possible to divide a larger whole number into a smaller whole number. The answer is a decimal fraction.

To divide 5 into 4, add a decimal point and a zero. Then divide.

As shown in the second example, adding several zeros is often a good way to eliminate a remainder in a division problem.

Examples:

$$5\overline{)4} \qquad 8\overline{)5}$$

$$\begin{array}{r} .8 \\ 5\overline{)4.0} \\ \underline{40} \\ 0 \end{array} \qquad \begin{array}{r} .625 \\ 8\overline{)5.000} \\ \underline{48} \\ 20 \\ \underline{16} \\ 40 \\ \underline{40} \\ 0 \end{array}$$

Use zeros to divide in each problem below. Add enough zeros in each problem so that division ends with no remainder.

Skill Builders

10.

$$2\overline{)1} \;\rightarrow\; 2\overline{)1.0} \qquad\qquad 4\overline{)7} \;\rightarrow\; 4\overline{)7.00} \qquad\qquad 8\overline{)3} \;\rightarrow\; 8\overline{)3.000}$$

11. $5\overline{)2}$ $\qquad\qquad$ $4\overline{)1}$ $\qquad\qquad$ $8\overline{)2}$ $\qquad\qquad$ $4\overline{)3}$ $\qquad\qquad$ $5\overline{)6}$

12. $8\overline{)3}$ $\qquad\qquad$ $4\overline{)7}$ $\qquad\qquad$ $8\overline{)7}$ $\qquad\qquad$ $4\overline{)5}$ $\qquad\qquad$ $16\overline{)5}$

Dividing a Decimal by a Decimal

To divide a decimal by a decimal, the first step is to change the divisor to a whole number. To do this you move the decimal point of the divisor (the number outside the division sign) to the right as far as you can. **Then you move the decimal point in the dividend an equal number of places**.

Example: Divide 2.832 by .03

Step 1. Set up the problem for long division.

Make the divisor (.03) a whole number. Do this by moving the decimal point two places to the right. Then move the decimal point in the dividend (2.832) two places to the right.

$$.03\overline{)2.832}$$

$$\begin{array}{r} 94.4 \\ 03.\overline{)283.2} \\ \underline{27} \\ 13 \\ \underline{12} \\ 12 \\ \underline{12} \\ 0 \end{array}$$

Step 2. Now divide the whole number (3) into the new decimal number 283.2.

Answer: 94.4

Moving the decimal point an equal number of places in both divisor and dividend makes sure that the decimal point appears in the correct place in the answer.

Solve each division problem below. Complete the partially worked Skill Builders by finishing any math and placing a decimal point in each answer.

Skill Builders

1.

$$.04\overline{)1.68} \quad \begin{array}{r}42\end{array}$$

$$.05\overline{)\,.32\,5} \quad \begin{array}{r}6.5 \\ \underline{30} \\ 2\,5 \\ \underline{2\,5}\end{array}$$

$$1.4\overline{)18.2} \quad \begin{array}{r}1\,3 \\ \underline{14} \\ 4\,2 \\ \underline{4\,2}\end{array}$$

$$.26\overline{)\,.806}$$

$$2.1\overline{)\,.441}$$

2. $.02\overline{)1.6}$ $.12\overline{)\,.48}$ $3.1\overline{)\,.093}$ $.004\overline{)8.408}$ $.07\overline{)49.14}$

3. $.04\overline{)3.108}$ $.006\overline{)91.44}$ $2.3\overline{)\,.529}$ $.07\overline{)355.6}$ $1.2\overline{)27.84}$

Using Zeros as Place Holders

It is often necessary to add one or more zeros to the dividend before it is possible to move the decimal point to the right.

In example 1 at right, we must move the decimal point two places in order to make the divisor (.04) a whole number. To move the decimal point in the dividend (2.8) two places, we must add one zero.

In example 2, three zeros must be added.

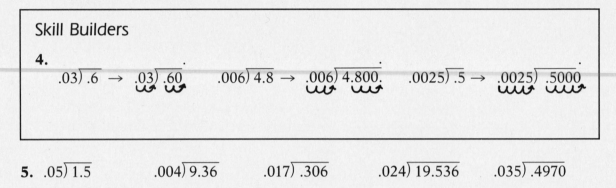

Example 1:

$.04\overline{)2.8}$

$$\overset{70.}{.04\overline{)2.80}}$$

Answer: 70

Example 2:

$.0006\overline{)3.6}$

$$\overset{6000.}{.0006\overline{)3.6000}}$$

Answer: 6,000

Skill Builders

4.

$.03\overline{)\,.6} \rightarrow .03\overline{)\,.60}$ $.006\overline{)4.8} \rightarrow .006\overline{)4.800}$ $.0025\overline{)\,.5} \rightarrow .0025\overline{)\,.5000}$

---

5. $.05\overline{)1.5}$ $.004\overline{)9.36}$ $.017\overline{)\,.306}$ $.024\overline{)19.536}$ $.035\overline{)\,.4970}$

Dividing Whole Numbers by Decimals

To divide a whole number by a decimal, add a decimal point to the whole number. Then add zeros as needed, move the decimal point, and divide.

To divide .08 into 4, the first step is to add a decimal point to 4. Then the decimal point in both .08 and 4. are moved to the right two places, and 8 is divided into 400.

Example:

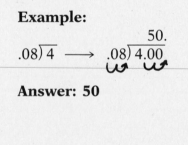

$.08\overline{)4} \longrightarrow \overset{50.}{.08\overline{)4.00}}$

Answer: 50

6. $.06\overline{)12}$ $1.5\overline{)30}$ $.003\overline{)6}$ $2.5\overline{)50}$ $.04\overline{)24}$

Dividing by 10, 100, or 1,000

There are three shortcuts we can use when dividing decimal numbers by 10, 100 or 1,000.

1. **To divide a decimal by 10, move the decimal point one place to the left.**

   Example 1: Divide 3.4 by 10.
   $$3.4 \div 10 = 3.4$$
   $$= .34$$

   Example 2: Divide 5 by 10.
   $$5 \div 10 = 5.$$
   $$= .5$$

2. **To divide a decimal by 100, move the decimal point two places to the left.**

   Example 3: Divide 2.5 by 100.
   $$2.5 \div 100$$
   $$= 02.5 = .025$$

   Example 4: Divide .34 by 100.
   $$.34 \div 100 =$$
   $$00.34 = .0034$$

   Notice in examples 3 and 4 that zeros have to be added to the left side of each number. These zeros hold places so that we can move the decimal point the correct number of places. These added zeros become part of the answer.

3. **To divide a decimal by 1,000, move the decimal point three places to the left.**

   Example 5: Divide 23.56 by 1,000
   $$23.56 \div 1,000 = .023\,56 = .02356$$

   To remember these three rules, notice that the decimal point is always moved the same number of places as there are 0s in the dividing number.

Using the shortcuts, divide each number below. The first one in each row is done as an example.

1. $5.6 \div 10 = .56$ $7.5 \div 10 =$ $23.5 \div 10 =$ $525.2 \div 10 =$

2. $37.2 \div 100 = .372$ $63.9 \div 100 =$ $3.5 \div 100 =$ $.021 \div 100 =$

3. $126 \div 10 = 12.6$ $15 \div 100 =$ $375 \div 100 =$ $8 \div 100 =$

4. $47.5 \div 1,000 = .0475$ $12 \div 1,000 =$ $.002 \div 1,000 =$

57

Solving Multiplication and Division Word Problems

In problems 1 through 4 underline a key word that suggests either multiplication or division. Then solve each problem.

1. The price of a 4.5-pound package of chicken is $6.03. At this rate, what is the cost of chicken per pound?

2. Sam earns "time and a half" for each hour of overtime he works. His overtime pay rate is found by multiplying his regular pay rate by 1.5. What is Sam's overtime rate when his regular hourly pay is $5.76?

3. While traveling through Vermont, Alice drove 315.9 miles on 13.5 gallons of gas. Knowing this, figure out how many miles Alice can drive on a single gallon of gas.

4. Orsen and three friends agreed to equally share the cost of renting camping gear for the weekend. If the total cost of the gear was $29.92, how much did Orsen need to pay?

Problems 5 through 8 contain extra information. In each problem, circle only the necessary information and then solve the problem.

5. A medium-size shipping box is packed with 144 cans of corn and weighs 216 pounds. Each can weighs 1.5 pounds. Packed with the same corn, a large-size box weighs 318 pounds. How many cans does this large-size box contain?

6. Myrna works part-time for Betty's Typing Service. Betty charges customers $8.50 per hour for typing. Of this amount, Myrna is paid $5.28. How much did Myrna make last week if she worked for 21.5 hours?

7. In the metric system, weight is measured in grams, a unit much smaller than the ounce. It takes 28.4 grams to equal one ounce and 453.6 grams to equal one pound. Use this information to determine the weight in grams of a 6-ounce can of tuna fish.

8. Fred designed and built an oak wall unit for less than $200. In it he placed 7 oak boards to use as bookshelves. Each board was 4.25 feet long and cost $6.97. At this rate, how much did Fred pay for each foot of oak shelving?

In problems 9 through 12, look at the list or drawing to the right of each problem to find necessary information. Then solve each problem.

9. The sign at the U-Park Garage lists the parking rates as shown at right. How much would it cost a tourist to park her car for 6 hours at this garage?

U-Park Garage
Hourly rate: $1.25
Daily rate: $8.75
Monthly rate: $145.00

10. At Cut-Rate Lumber Products, the price of a fir board depends upon the length you buy. A partial list of prices is shown at right. According to this list, what price do you pay per foot for a 14-foot-long board?

Cut-Rate Lumber
17-foot length $17.85
14-foot length $13.30
9-foot length $ 6.57

11. As shown in her drawing at right, a machinist placed 3 nuts of equal width on a bolt to use as a spacer. What length of the 1.5-inch bolt is taken up by the 3 nuts?

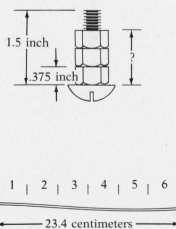

12. For his jewelry-making business, Jason uses small pieces of solid silver wire. As shown at right, he wants to divide this long wire into 6 equal lengths. How long should he cut each piece?

| 1 | 2 | 3 | 4 | 5 | 6 |

←————— 23.4 centimeters —————→

Solving Multi-Step Word Problems

Problems 1 through 6 are multi-step problems. In each a solution sentence is written and the missing information is underlined. Solve each problem after first finding the value of the missing information. The first one is started for you.

1. In March Lena bought a new television for $469.88. She made a down payment of $125.00. Because she agreed to pay off the balance in 6 equal monthly payments, she paid no interest charges. Compute the amount of Lena's monthly payment.

   monthly payment = <u>amount of balance</u> divided by 6
   = (cost of TV − down payment) ÷ 6

2. David lives 1.4 miles from school. He walks to school each morning and home each night five days a week. How many miles does David walk each week going to and from school?

   miles walked each week = 5 times <u>total miles walked each day</u>

3. How much change would you expect to receive if you bought 5.5 pounds of nails on sale for $.97 per pound and you paid with a ten-dollar bill?

   change = $10.00 minus <u>total cost of nails</u>

4. During last week, Ester worked 40 hours at her regular hourly rate of $6.50 per hour. She also worked 6.4 hours of overtime last Saturday. Her overtime rate is $9.75 per hour. How much did Ester earn in all last week?

   total earnings = <u>regular earnings</u> plus <u>overtime earnings</u>

5. Jannie jogs 1.7 miles each Monday, Wednesday, and Friday. On Tuesday and Thursday she jogs 2.4 miles. How many total miles does Jannie jog each week?

   total miles jogged = <u>miles jogged on M, W, and F</u> plus <u>miles jogged on Tu and Th</u>

6. As a decimal fraction, one-sixteenth inch is equal to .0625 inch. Russ, a machinist, uses washers that are one-sixteenth inch thick. How much room will be left on a 1-inch-long bolt if he places 5 of these washers on the bolt?

   room left = 1 inch minus <u>thickness of 5 washers</u>

Washer

top view

side view .0625 in.

Solve problems 7 through 12. You may find it helpful in each problem to first write a solution sentence to help you identify missing information.

7. At a Thanksgiving Sale, Jerry bought a 3.7-pound package of chicken priced at $.70 per pound, and a 4.6-pound roast priced at $1.75 per pound. What amount did Jerry pay for these two packages of meat?

8. Norm's Road Construction Company is fixing potholes along Airport Drive. The crew can average fixing about .42 miles of road per hour. Working 10 hours each day, how many days will it take them to repair the 21-mile-long road?

9. Mark bought a 23.5-pound sack of potatoes marked at $.24 per pound. If he pays by check, for how much should he write the check if he wants to get $10 back in change?

10. In the metric system, road distance is measured in kilometers. A kilometer is shorter than a mile. In fact, 1 kilometer is equal to .62 mile. Giving your answer in miles, how much shorter is 10 kilometers than 10 miles?

|←—————— 10 miles ——————→|

|←——— 10 kilometers ——→|

|←——— ? miles ———→|←— ? miles —→|

In problems 11 and 12, circle the arithmetic expression that will give the correct answer to each question. You do not need to solve these two problems.

11. As a decimal, three and one-eighth pounds is written as 3.125 pounds. How much heavier than 14 pounds is a group of five 3.125-pound weights?

a) 14 × 5 − 3.125
b) 5 × 3.125 − 14
c) 14 − 5 × 3.125
d) 14 (5 + 3.125)

12. Last month, Kate worked 176 hours at her regular pay rate of $5.00 per hour. She also worked 16.5 hours of overtime at a pay rate of $7.50 per hour. How much total salary did Kate earn last month?

a) 176 × 5 − 16.5 × 7.5
b) 176 × 5 + 16.5 × 7.5
c) 176 (5 + 7.5)
d) (176 + 16.5) × (7.5 + 5)

Rounding an answer is useful in a wide variety of decimal word problems. In problems 13 through 17, round each answer as indicated.

13. Gasoline prices are always given in dollars, cents, and parts of a cent. For example, a price of $1.299 per gallon means $1.29 and nine tenths cent per gallon. Rounding to the nearest penny, how much would 18.7 gallons of gas cost at this price?

14. According to the blueprint, Jan is supposed to cut the bushing to a thickness of .2573 inch. The rough-cut bushing starts at .27 inch. How much does Jan need to cut off the bushing to meet the specifications of the blueprint? Express your answer to the nearest thousandth of an inch.

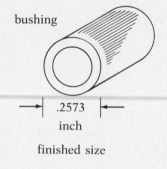

bushing

.2573 inch

finished size

15. A hand-held calculator expresses numbers smaller than 1 as decimal fractions. And a calculator always rounds a decimal fraction by simply dropping extra digits. Suppose you divide 2 by 7 on a calculator that rounds to 4 decimal places. What answer will show on the calculator display window?

16. A **repeating decimal** occurs when you divide two numbers and can't get rid of the remainder by continued dividing. As shown at right, 5 divided by 6 is an example. Continuing dividing by adding more zeros to the right of the decimal point does not get rid of the remainder.

```
      .8333......
6) 5.0000......
   4 8
     20
     18
     20
     18
     20
     18
      2
```

To see a second example, divide 2 by 3. Round your answer to the nearest thousandths place.

Questions 17, 18, and 19 refer to the following story.

Joyce Hernandez manages a service station. Before June 1 she sold unleaded gas at a price of $1.039 per gallon. Then, because of a drop in wholesale prices, she lowered the pump price of unleaded by $.019 per gallon.

17. After June 1, what price did Joyce charge per gallon for unleaded gas?

18. At this new price, how much does 15.4 gallons cost?

19. At the lower price, how many gallons of unleaded can you buy for $16.00? Express your answer to the nearest tenth gallon.

Questions 20 through 22 refer to the story below.

Jamie wants to decide which of two day-care centers would be the least expensive for her child to attend. Because of her work schedule, Jamie will need to send her child to the center 18 days each month, and the child will need lunch each day.

School for Kids charges by the month. It charges $137.50 for each child enrolled in the program. There is no refund if you don't bring your child every day, and lunch is provided free.

Little Bunnies Center charges by the day. It charges $7.00 per child per day. Lunch at Little Bunnies is extra. A $1.00 fee is charged for each lunch for each child.

20. How much would it cost Jamie to send her child to School for Kids for 18 days each month?

21. What would Jamie have to pay each day if her child went to Little Bunnies Center?

22. How much would it cost Jamie to send her child to Little Bunnies Center for 18 days each month?

Decimal Skills Review

As you complete this chapter on decimals, it is a good idea to briefly review the main decimal computation skills. Work each problem below as carefully as you can.

Comparing Values of Decimal Fractions:

Review pages 34 through 39.

In each group of decimal fractions below, circle the two that are equal in value.

1. .004, .040, .04   .503, .053, .5030   .390, .039, .39

2. .0520, .052, .0052   .100, .010, .10   .0308, .03080, .3080

Rounding Decimal Fractions: Review pages 40 through 41.

Round each decimal fraction below to the nearest 100th place.

3. .452   .093   .106   .0075   .4901

Adding and Subtracting Decimals:

Review pages 42 through 45.

When you are adding or subtracting decimals, line up the decimal points as your first step. Then add or subtract in the same way you do with whole numbers. Add zeros when necessary.

4. .6 + .8   1.4 + .9   3.24 + 2   .05 + 2.1   5 + .7

5. .9 − .5   4.8 − .9   6.48 − 4   .46 − .3   8 − 4.7

6. 2 + 3.5 + .81   5.67 + 2.3 + 3   12.4 + 6 + .57

7. $7.12 − $.89 $25.00 − $15.69 $8.05 − 97¢ $157 − $64.48

8. $8.46 + $1.83 $5.00 + $2.79 $12.83 + 75¢ $67.28 + $13

Multiplying Decimals: Review pages 48 through 51.

When you are multiplying decimals, the number of decimal places in the answer must equal the total number of decimal places in the problem.

9.
$$.9 \times 7 \qquad .7 \times 4 \qquad 2.6 \times 5 \qquad 3.8 \times 7 \qquad 12 \times .6 \qquad 14 \times .2$$

10.
$$2.4 \times 11 \qquad 7.5 \times 10 \qquad 26 \times 2.1 \qquad 100 \times 4.5 \qquad 5.67 \times 100 \qquad 3.7 \times 10$$

11.
$$.8 \times .7 \qquad .9 \times .5 \qquad 1.5 \times .6 \qquad .85 \times .7 \qquad .45 \times .8 \qquad .054 \times .8$$

12.
$$5.8 \times .37 \qquad 8.2 \times .48 \qquad .56 \times .28 \qquad 12.6 \times 2.8 \qquad .37 \times 1.4 \qquad 1.66 \times .63$$

13. $3.47 \times 1{,}000 =$ $\qquad$ $.28 \times 100 =$ $\qquad$ $5.9 \times 10 =$

Dividing Decimals: Review pages 52 through 57.

When you are dividing decimals, your first step is to change the divisor (the number *outside* the bracket) to a whole number. Do this by moving the decimal points of both dividend and divisor an equal number of places.

14. $6\overline{)3.06}$ $\qquad$ $7\overline{)2.87}$ $\qquad$ $5\overline{)25.10}$ $\qquad$ $12\overline{)4.824}$ $\qquad$ $9\overline{)1.827}$

15. $7\overline{)2.52}$ $\qquad$ $8\overline{).096}$ $\qquad$ $3\overline{).117}$ $\qquad$ $1.2\overline{)2.472}$ $\qquad$ $.7\overline{).385}$

16. $3.5\overline{)1.470}$ $\qquad$ $.023\overline{)124.2}$ $\qquad$ $.004\overline{)90}$ $\qquad$ $4 \div 1{,}000 =$

4
Common Fraction Skills

A common fraction stands for one or more parts of a whole. A common fraction is written as a top number above a bottom number. The top number is called the **numerator**, and the bottom number is called the **denominator**. Most often, a common fraction is just called a "fraction."

$\frac{5}{8}$ ← numerator   The **numerator** tells how many parts you have.

    ← denominator   The **denominator** tells how many parts one whole is divided into.

Proper Fractions

The fraction you see most often is a **proper fraction**. In a proper fraction, the top number is always smaller than the bottom number. The value of a proper fraction is always less than 1.

Example 1: $\frac{5}{8}$ A pie is cut into 8 equal pieces. Only 5 of those pieces remain.

$\frac{5}{8}$ of the pie remains. $\frac{3}{8}$ is gone.

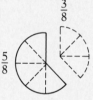

Improper Fractions

In an **improper fraction**, the top number is either the same as or larger than the bottom number. The value of an improper fraction is either 1 or larger than 1.

Example 2: $\frac{3}{3}$ The top 3 stands for the number of pieces you have. But 3 is also the number of pieces in 1 whole. Thus, the value of $\frac{3}{3}$ is 1.

Example 3: $\frac{7}{4}$ One whole is divided into 4 equal pieces. But, you have 7 pieces. This is 3 pieces more than 1 whole. Therefore, the value of $\frac{7}{4}$ is larger than 1.

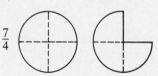

66

Mixed Numbers

A **mixed number** consists of a fraction written next to a whole number. A mixed number is understood to be the sum of the whole number and the fraction.

Example 4: $2\frac{1}{2}$ The whole number 2 stands for 2 whole objects. In addition to the 2 objects, there is $\frac{1}{2}$ object more.

$2\frac{1}{2}$ ◯ ◯ ◗

Each whole figure below is divided into equal parts, some shaded and some white. On the first line, write the proper fraction that represents the shaded part. On the second line, write the fraction that represents the unshaded (white) part.

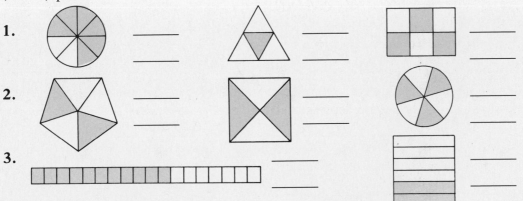

1. _____   _____   _____
    _____   _____   _____

2. _____   _____   _____
    _____   _____   _____

3. _____   _____
    _____   _____

Write *p*, *i*, or *m*, to indicate whether each of the following numbers is a proper fraction, an improper fraction, or a mixed number.

4. $\frac{4}{5}$ _____ $\frac{9}{8}$ _____ $4\frac{1}{3}$ _____ $\frac{12}{10}$ _____ $2\frac{1}{2}$ _____

5. $\frac{7}{7}$ _____ $\frac{5}{4}$ _____ $\frac{7}{8}$ _____ $\frac{12}{12}$ _____ $4\frac{5}{12}$ _____

Beside each drawing, write *p*, *i*, or *m*, to indicate which type of fraction or mixed number each drawing represents. Consider a complete circle as 1 whole.

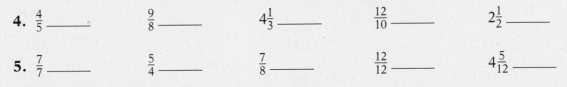

6. _____   _____   _____

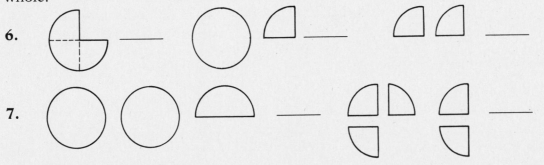

7. _____   _____   _____

Simplifying Fractions

On these next two pages we'll look at two ways that are commonly used to simplify the writing of fractions. You'll use both of these throughout your work with fractions.

Reducing Fractions

To reduce a fraction is to rewrite it using smaller numbers. Reducing does not change the value of a fraction. It simply replaces one fraction with another of equal value called an **equivalent fraction**. Here are three examples of equivalent fractions:

Equivalent Fractions

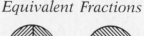

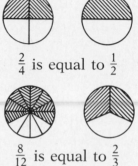

$\frac{2}{4}$ is equal to $\frac{1}{2}$

$\frac{8}{12}$ is equal to $\frac{2}{3}$

a) $\frac{2}{4} = \frac{1}{2}$ b) $\frac{8}{12} = \frac{2}{3}$ c) $\frac{6}{10} = \frac{3}{5}$

When a fraction is in its simplest form — the smallest numbers possible — it is said to be **reduced to lowest terms**.

To reduce a fraction to lowest terms, divide both numerator and denominator by the largest whole number that divides evenly into each.

Example 1: Reduce $\frac{3}{9}$ to lowest terms.

Divide both numerator (3) and denominator (9) by 3.

$$\frac{3 \div 3}{9 \div 3} = \frac{1}{3}$$

Answer: $\frac{1}{3}$

Example 2: Reduce $\frac{12}{20}$ to lowest terms.

Divide both 12 and 20 by 2. Divide by 2 again.

$$\frac{12 \div 2}{20 \div 2} = \frac{6}{10} = \frac{3}{5}$$

Answer: $\frac{3}{5}$

After you reduce a fraction, see if it can be reduced further. Notice that in Example 2, $\frac{12}{20}$ was reduced twice. An alternative would have been to divide both the numerator and denominator by 4. Often it is difficult to see the largest whole number to divide by the first time you reduce, so you have to reduce again.

Reduce each fraction to lowest terms.

1. $\frac{4}{12} =$ $\frac{6}{8} =$ $\frac{4}{6} =$ $\frac{6}{9} =$ $\frac{12}{16} =$

2. $\frac{9}{15} =$ $\frac{15}{25} =$ $\frac{24}{30} =$ $\frac{14}{21} =$ $\frac{10}{20} =$

3. $\frac{3}{9} =$ $\frac{2}{8} =$ $\frac{30}{35} =$ $\frac{14}{28} =$ $\frac{12}{15} =$

Changing Improper Fractions to Mixed Numbers

Many answers to fraction addition and multiplication problems are first written as improper fractions. To write these answers in simplest form, we change them to mixed numbers.

To change an improper fraction to a mixed number, divide the denominator into the numerator.

Example 1: Change $\frac{13}{2}$ to a mixed number.

Step 1. Divide 2 into 13:
$$\begin{array}{r} 6 \text{ r}1 \\ 2\overline{)13} \\ \underline{12} \\ 1 \end{array}$$

Step 2. Write the remainder (1) over the divisor (2) to form the proper fraction part of the answer.

Answer: $6\frac{1}{2}$

Example 2: Change $\frac{12}{8}$ to a mixed number.

Step 1. Divide 8 into 12:
$$\begin{array}{r} 1 \text{ r}4 \\ 8\overline{)12} \\ \underline{8} \\ 4 \end{array}$$

$1 \text{ r}4 = 1\frac{4}{8}$

Step 2. Reduce the proper fraction.
$$\frac{4 \div 4}{8 \div 4} = \frac{1}{2}$$

Answer: $1\frac{4}{8} = 1\frac{1}{2}$

On the first line below each picture, write an improper fraction that stands for the amount shown. On the second line, write this same amount as a mixed number.

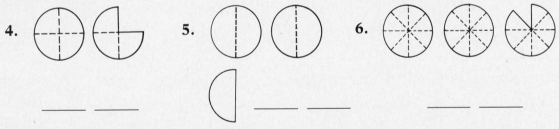

4.   5.   6.

_____ _____   _____ _____   _____ _____

Change each improper fraction below to a mixed number. Reduce proper fractions.

7. $\frac{21}{10} =$ $\frac{5}{2} =$ $\frac{17}{9} =$ $\frac{19}{11} =$ $\frac{7}{4} =$

8. $\frac{14}{8} =$ $\frac{20}{15} =$ $\frac{9}{6} =$ $\frac{22}{4} =$ $\frac{23}{5} =$

9. At Leo's Pizza by the Slice, Leo cuts whole pizzas into 8 slices. Monday night there were 21 slices left over. How many pizzas is this? (Hint: $\frac{21}{8}$ pizzas are left over.)

10. Jenny has $\frac{21}{2}$ yards of dress material left on a roll. Does she have enough to sell this as a "half-roll" if a half-roll is supposed to contain at least 10 yards of material?

69

Raising Fractions to Higher Terms

To raise a fraction to higher terms is to rewrite the fraction using larger numbers. Raising a fraction to higher terms is the opposite of reducing a fraction. In both cases, the fractions remain *equivalent*, or equal in value.

To raise a fraction to higher terms, multiply both numerator and denominator by the same number.

Example 1: Write $\frac{2}{5}$ as a fraction that has 15 as a denominator.

Step 1. Ask yourself, "What number times 5 equals 15?" $\qquad \frac{2}{5} = \frac{?}{15}$

You must multiply 2 by the same number to find the missing numerator.

Step 2. To answer the question, divide 5 into 15: $\qquad \frac{2}{5} \searrow\nearrow \frac{6}{15}$
$15 \div 5 = 3$

Step 3. Multiply 2 by 3: $\qquad\qquad$ Divide 5 into 15 and get 3.
$2 \times 3 = 6 \qquad\qquad$ Multiply 2 by 3 to get the
missing numerator, 6.

Answer: $\frac{2}{5} = \frac{6}{15}$

In examples 2 and 3, we'll list just the computation steps. Remembering the "backward Z" formed by the arrows helps many students remember these steps.

Example 2: Rewrite $\frac{1}{3}$ as twelfths. $\qquad$ **Example 3:** $\frac{3}{4} = \frac{?}{16}$

Step 1. Divide 3 into 12: $\qquad\qquad$ *Step 1.* Divide 4 into 16:
$12 \div 3 = 4 \qquad\qquad\qquad\qquad 16 \div 4 = 4$

$\frac{1}{3} \searrow\nearrow \frac{}{12} \qquad\qquad\qquad \frac{3}{4} \searrow\nearrow \frac{}{16}$

Step 2. Multiply 1 by 4: $\qquad\qquad$ *Step 2.* Multiply 3 by 4:
$1 \times 4 = 4 \qquad\qquad\qquad\qquad 3 \times 4 = 12$

Answer: $\frac{1}{3} = \frac{4}{12} \qquad\qquad$ **Answer:** $\frac{3}{4} = \frac{12}{16}$

Raise each fraction to higher terms by writing the numerator of each new fraction.

1. $\frac{1}{2} = \frac{}{8}$ $\qquad$ $\frac{1}{3} = \frac{}{6}$ $\qquad$ $\frac{1}{4} = \frac{}{8}$ $\qquad$ $\frac{2}{3} = \frac{}{9}$ $\qquad$ $\frac{3}{4} = \frac{}{12}$

2. $\frac{2}{5} = \frac{}{10}$ $\qquad$ $\frac{3}{4} = \frac{}{8}$ $\qquad$ $\frac{1}{3} = \frac{}{12}$ $\qquad$ $\frac{5}{7} = \frac{}{14}$ $\qquad$ $\frac{1}{2} = \frac{}{10}$

3. $\frac{1}{3} = \frac{}{15}$ $\qquad$ $\frac{2}{5} = \frac{}{30}$ $\qquad$ $\frac{5}{14} = \frac{}{42}$ $\qquad$ $\frac{11}{12} = \frac{}{24}$ $\qquad$ $\frac{4}{7} = \frac{}{42}$

Comparing Common Fractions

Common fractions are easily compared when they have the same denominator. For example, $\frac{3}{4}$ is larger than $\frac{2}{4}$ because 3 is larger than 2. The rules for comparing fractions are easy to follow:

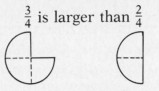

$\frac{3}{4}$ is larger than $\frac{2}{4}$

Rules for Comparing Common Fractions

1. Write each fraction so that all fractions have the same (common) denominator. This usually means raising one or more fractions to higher terms.

2. Compare the numerators of the like fractions that result from rule 1.

Example: Of the following three fractions, which is the largest and which is the smallest: $\frac{1}{4}$, $\frac{3}{16}$, and $\frac{5}{32}$?

Step 1. To compare, we can choose 32 as a common denominator and rewrite $\frac{1}{4}$ and $\frac{3}{16}$ as fractions with a denominator of 32.

$$\frac{1}{4} = \frac{8}{32} \qquad \frac{3}{16} = \frac{6}{32} \qquad \frac{5}{32}$$

Step 2. Compare the numerators of the like fractions: 5, 6, and 8. From smallest to largest, the fractions are:

smallest: $\frac{5}{32}$ middle value: $\frac{6}{32}$ $(\frac{3}{16})$ largest: $\frac{8}{32}$ $(\frac{1}{4})$

Answer: $\frac{5}{32}$ **is the smallest.** $\frac{1}{4}$ **is the largest.**

In each pair below, circle the larger fraction. Use the larger denominator in each pair as the common denominator with which to write like fractions.

1. $\frac{2}{3}$ or $\frac{7}{12}$ $\frac{3}{4}$ or $\frac{10}{12}$ $\frac{5}{8}$ or $\frac{1}{2}$ $\frac{7}{8}$ or $\frac{13}{16}$

2. $\frac{1}{4}$ or $\frac{3}{8}$ $\frac{4}{9}$ or $\frac{1}{3}$ $\frac{3}{10}$ or $\frac{2}{5}$ $\frac{6}{7}$ or $\frac{11}{14}$

Arrange each group of three fractions in order. Write the smallest to the left and the largest to the right. Use the largest denominator as the common denominator.

3. $\frac{1}{3}$, $\frac{5}{12}$, $\frac{1}{4}$ $\frac{2}{3}$, $\frac{7}{9}$, $\frac{1}{3}$ $\frac{7}{24}$, $\frac{3}{8}$, $\frac{1}{3}$ $\frac{6}{14}$, $\frac{4}{7}$, $\frac{1}{2}$

Finding What Fraction a Part Is of a Whole

Problems of measurement often require you to write one amount as a fraction of a larger unit. We'll discuss this skill on this page.

Example: There are 36 inches in a yard. What fraction of a yard is 28 inches?

Step 1. Write a fraction by placing the $\frac{28}{36}$ part (28) over the whole (36).

Step 2. Reduce this fraction: $\frac{28}{36} = \frac{28 \div 4}{36 \div 4} = \frac{7}{9}$

Answer: **28 inches is $\frac{7}{9}$ of a yard.**

Express each amount below as a fraction of the larger unit indicated. Write each fraction in its most reduced form.

1. 1 yard = 36 inches

6 inches = ____ yard

24 inches = ____ yard

30 inches = ____ yard

2. 1 foot = 12 inches

6 inches = ____ foot

8 inches = ____ foot

10 inches = ____ foot

3. 1 hour = 60 minutes

8 minutes = ____ hour

20 minutes = ____ hour

45 minutes = ____ hour

4. 1 year = 52 weeks

16 weeks = ____ year

32 weeks = ____ year

48 weeks = ____ year

5. 1 pound = 16 ounces

8 ounces = ____ pound

10 ounces = ____ pound

14 ounces = ____ pound

6. 1 year = 12 months

3 months = ____ year

4 months = ____ year

10 months = ____ year

7. Al makes $1,500 each month on his job as a salesman. If he pays $375 each month for rent, what fraction of his salary is his rent?

8. Lee read 240 pages of a 1,000-page novel. What fraction of the novel has she read?

Adding Like Fractions

Fractions that have the same denominator are called **like fractions**. For example, $\frac{3}{8}$ and $\frac{5}{8}$ are like fractions, while $\frac{3}{8}$ and $\frac{5}{16}$ are not.

To add like fractions, add the numerators and place the sum over the denominator. Reduce the answer to lowest terms.

Example: Add $\frac{7}{16}$ and $\frac{5}{16}$.

Step 1. Add the numerators.
$7 + 5 = 12$

Step 2. Place the sum 12 over the denominator 16.

Step 3. Reduce $\frac{12}{16}$ to lowest terms. Divide both 12 and 16 by 4.

$$\frac{12 \div 4}{16 \div 4} = \frac{3}{4}$$

$$\begin{array}{r} \frac{7}{16} \\ + \frac{5}{16} \\ \hline \frac{12}{16} = \frac{3}{4} \end{array}$$

Think of a bar divided into 16 equal parts. Now add as shown.

$$\frac{7}{16} + \frac{5}{16} = \frac{7+5}{16}$$

$$\frac{12}{16}$$

Note: *You do not add denominators.*

Answer: $\frac{3}{4}$

Add. Write each answer in lowest terms. The first problem in each row is done as an example.

Fractions that Add to Less than 1

1. $\begin{array}{r} \frac{2}{4} \\ + \frac{1}{4} \\ \hline \frac{3}{4} \end{array}$
$\begin{array}{r} \frac{1}{3} \\ + \frac{1}{3} \\ \hline \end{array}$
$\begin{array}{r} \frac{3}{5} \\ + \frac{1}{5} \\ \hline \end{array}$
$\begin{array}{r} \frac{5}{8} \\ + \frac{2}{8} \\ \hline \end{array}$
$\begin{array}{r} \frac{7}{12} \\ + \frac{4}{12} \\ \hline \end{array}$

2. $\frac{5}{8} + \frac{1}{8} = \frac{6}{8}$ $\qquad$ $\frac{1}{4} + \frac{1}{4}$ $\qquad$ $\frac{5}{9} + \frac{1}{9}$ $\qquad$ $\frac{3}{8} + \frac{3}{8}$ $\qquad$ $\frac{9}{16} + \frac{5}{16}$

$\frac{6 \div 2}{8 \div 2} = \frac{3}{4}$

3. $\begin{array}{r} \frac{7}{16} \\ \frac{4}{16} \\ + \frac{3}{16} \\ \hline \frac{14}{16} \end{array}$
$\begin{array}{r} \frac{3}{8} \\ \frac{2}{8} \\ + \frac{1}{8} \\ \hline \end{array}$
$\begin{array}{r} \frac{5}{12} \\ \frac{3}{12} \\ + \frac{2}{12} \\ \hline \end{array}$
$\begin{array}{r} \frac{5}{9} \\ \frac{1}{9} \\ + \frac{2}{9} \\ \hline \end{array}$
$\begin{array}{r} \frac{9}{14} \\ \frac{3}{14} \\ + \frac{1}{14} \\ \hline \end{array}$

$\frac{14 \div 2}{16 \div 2} = \frac{7}{8}$

Fractions that Add to a Whole Number

A fraction that has the same top and bottom number is always equal to 1.

As shown at right, when fraction addition gives an answer with the same numerator and denominator, write the answer as 1.

Example:

$$\begin{array}{r} \frac{7}{10} \\ +\frac{3}{10} \\ \hline \frac{10}{10} = 1 \end{array}$$

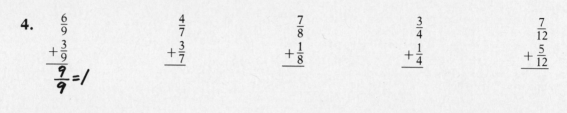

4.

$$\begin{array}{r} \frac{6}{9} \\ +\frac{3}{9} \\ \hline \frac{9}{9} = 1 \end{array}$$

$$\begin{array}{r} \frac{4}{7} \\ +\frac{3}{7} \\ \hline \end{array}$$

$$\begin{array}{r} \frac{7}{8} \\ +\frac{1}{8} \\ \hline \end{array}$$

$$\begin{array}{r} \frac{3}{4} \\ +\frac{1}{4} \\ \hline \end{array}$$

$$\begin{array}{r} \frac{7}{12} \\ +\frac{5}{12} \\ \hline \end{array}$$

As the following problems show, fractions may also add up to whole numbers larger than 1.

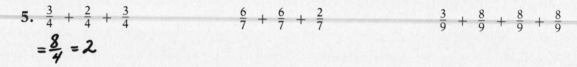

5. $\frac{3}{4} + \frac{2}{4} + \frac{3}{4}$ $\frac{6}{7} + \frac{6}{7} + \frac{2}{7}$ $\frac{3}{9} + \frac{8}{9} + \frac{8}{9} + \frac{8}{9}$

   $= \frac{8}{4} = 2$

Fractions that Add to More than 1

When fraction addition gives an answer that is an **improper fraction**, change the answer to a **mixed number**.

Reduce the proper fraction if possible.

Example:

$$\begin{array}{r} \frac{5}{7} \\ +\frac{4}{7} \\ \hline \frac{9}{7} = 1\frac{2}{7} \end{array}$$

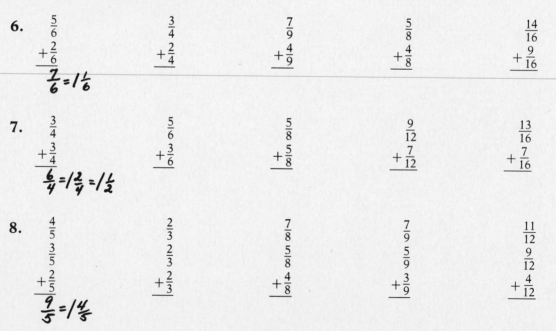

6.

$$\begin{array}{r} \frac{5}{6} \\ +\frac{2}{6} \\ \hline \frac{7}{6} = 1\frac{1}{6} \end{array}$$

$$\begin{array}{r} \frac{3}{4} \\ +\frac{2}{4} \\ \hline \end{array}$$

$$\begin{array}{r} \frac{7}{9} \\ +\frac{4}{9} \\ \hline \end{array}$$

$$\begin{array}{r} \frac{5}{8} \\ +\frac{4}{8} \\ \hline \end{array}$$

$$\begin{array}{r} \frac{14}{16} \\ +\frac{9}{16} \\ \hline \end{array}$$

7.

$$\begin{array}{r} \frac{3}{4} \\ +\frac{3}{4} \\ \hline \frac{6}{4} = 1\frac{2}{4} = 1\frac{1}{2} \end{array}$$

$$\begin{array}{r} \frac{5}{6} \\ +\frac{3}{6} \\ \hline \end{array}$$

$$\begin{array}{r} \frac{5}{8} \\ +\frac{5}{8} \\ \hline \end{array}$$

$$\begin{array}{r} \frac{9}{12} \\ +\frac{7}{12} \\ \hline \end{array}$$

$$\begin{array}{r} \frac{13}{16} \\ +\frac{7}{16} \\ \hline \end{array}$$

8.

$$\begin{array}{r} \frac{4}{5} \\ \frac{3}{5} \\ +\frac{2}{5} \\ \hline \frac{9}{5} = 1\frac{4}{5} \end{array}$$

$$\begin{array}{r} \frac{2}{3} \\ \frac{2}{3} \\ +\frac{2}{3} \\ \hline \end{array}$$

$$\begin{array}{r} \frac{7}{8} \\ \frac{5}{8} \\ +\frac{4}{8} \\ \hline \end{array}$$

$$\begin{array}{r} \frac{7}{9} \\ \frac{5}{9} \\ +\frac{3}{9} \\ \hline \end{array}$$

$$\begin{array}{r} \frac{11}{12} \\ \frac{9}{12} \\ +\frac{4}{12} \\ \hline \end{array}$$

Adding Fractions and Mixed Numbers

To add a fraction and a mixed number, first add the fractions alone. Then bring down the whole number.

If the sum of fractions is an **improper fraction**, change this sum to a **mixed number**. Add the whole number part of the mixed number to the whole number already in the answer.

Reduce the proper fraction if possible.

Example:
$$2\frac{4}{8}$$
$$+\ \frac{7}{8}$$
$$2\frac{11}{8} = 2 + 1\frac{3}{8}$$
$$= 3\frac{3}{8}$$

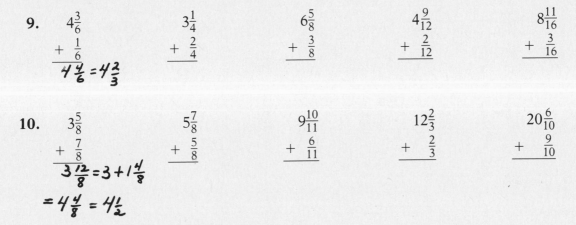

9.
$$4\frac{3}{6}$$
$$+\ \frac{1}{6}$$
$$4\frac{4}{6} = 4\frac{2}{3}$$

$$3\frac{1}{4}$$
$$+\ \frac{2}{4}$$

$$6\frac{5}{8}$$
$$+\ \frac{3}{8}$$

$$4\frac{9}{12}$$
$$+\ \frac{2}{12}$$

$$8\frac{11}{16}$$
$$+\ \frac{3}{16}$$

10.
$$3\frac{5}{8}$$
$$+\ \frac{7}{8}$$
$$3\frac{12}{8} = 3 + 1\frac{4}{8}$$
$$= 4\frac{4}{8} = 4\frac{1}{2}$$

$$5\frac{7}{8}$$
$$+\ \frac{5}{8}$$

$$9\frac{10}{11}$$
$$+\ \frac{6}{11}$$

$$12\frac{2}{3}$$
$$+\ \frac{2}{3}$$

$$20\frac{6}{10}$$
$$+\ \frac{9}{10}$$

Adding Mixed Numbers

To add mixed numbers, add the fractions and the whole numbers separately.

If the sum of fractions is an **improper fraction**, change this sum to a **mixed number**. Add the whole number part to the sum of whole numbers already in the answer.

Reduce the proper fraction if possible.

Example:
$$3\frac{7}{16}$$
$$+2\frac{11}{16}$$
$$5\frac{18}{16} = 5 + 1\frac{2}{16}$$
$$= 6\frac{2}{16} = 6\frac{1}{8}$$

11.
$$2\frac{3}{4}$$
$$+1\frac{1}{4}$$
$$3\frac{4}{4} = 3 + 1 = 4$$

$$4\frac{2}{4}$$
$$+3\frac{1}{4}$$

$$7\frac{2}{9}$$
$$+5\frac{4}{9}$$

$$8\frac{5}{12}$$
$$+6\frac{3}{12}$$

$$14\frac{13}{16}$$
$$+\ 7\frac{1}{16}$$

12. $\quad 4\frac{7}{8} + 2\frac{5}{8} \qquad 7\frac{8}{9} + 6\frac{7}{9} \qquad 8\frac{7}{8} + 2\frac{5}{8} \qquad 15\frac{11}{12} + 13\frac{9}{12} \qquad 25\frac{13}{16} + 19\frac{13}{16}$

$$= 6\frac{12}{8}$$
$$= 6 + 1\frac{4}{8}$$
$$= 7\frac{4}{8} = 7\frac{1}{2}$$

75

Subtracting Like Fractions

To subtract like fractions, subtract the numerators and place the difference over the denominator. Reduce the answer to lowest terms.

Example: Subtract $\frac{2}{9}$ from $\frac{8}{9}$.

Step 1. Subtract the numerators.
$8 - 2 = 6$

$$\begin{array}{r} \frac{8}{9} \\ -\frac{2}{9} \\ \hline \frac{6}{9} \end{array}$$

Think of a circle divided into 9 equal parts.

Step 2. Place the difference 6 over the denominator 9.

$\frac{8}{9}$ are there. Subtract $\frac{2}{9}$ and $\frac{8-2}{9} = \frac{6}{9}$ are left.

Step 3. Reduce $\frac{6}{9}$ to lowest terms. Divide both 6 and 9 by 3.

$$\frac{6 \div 3}{9 \div 3} = \frac{2}{3}$$

Note: *You do not subtract denominators.*

Answer: $\frac{2}{3}$

Subtract. Write each answer in lowest terms. The first problem in each row is done as an example.

Subtracting Two Fractions

1. $\begin{array}{r} \frac{6}{8} \\ -\frac{1}{8} \\ \hline \mathbf{\frac{5}{8}} \end{array}$ $\begin{array}{r} \frac{3}{4} \\ -\frac{2}{4} \\ \hline \end{array}$ $\begin{array}{r} \frac{7}{9} \\ -\frac{3}{9} \\ \hline \end{array}$ $\begin{array}{r} \frac{2}{3} \\ -\frac{1}{3} \\ \hline \end{array}$ $\begin{array}{r} \frac{9}{10} \\ -\frac{2}{10} \\ \hline \end{array}$

2. $\begin{array}{r} \frac{3}{4} \\ -\frac{1}{4} \\ \hline \mathbf{\frac{2}{4}=\frac{1}{2}} \end{array}$ $\begin{array}{r} \frac{8}{9} \\ -\frac{2}{9} \\ \hline \end{array}$ $\begin{array}{r} \frac{7}{8} \\ -\frac{3}{8} \\ \hline \end{array}$ $\begin{array}{r} \frac{11}{12} \\ -\frac{8}{12} \\ \hline \end{array}$ $\begin{array}{r} \frac{13}{16} \\ -\frac{3}{16} \\ \hline \end{array}$

Subtracting Fractions from Mixed Numbers

To subtract a fraction from a mixed number, first subtract the fractions. Then bring down the whole number. Reduce the fraction in the answer if possible.

3. $\begin{array}{r} 4\frac{5}{8} \\ -\frac{2}{8} \\ \hline \mathbf{4\frac{3}{8}} \end{array}$ $\begin{array}{r} 7\frac{8}{9} \\ -\frac{3}{9} \\ \hline \end{array}$ $\begin{array}{r} 5\frac{3}{5} \\ -\frac{2}{5} \\ \hline \end{array}$ $\begin{array}{r} 4\frac{2}{3} \\ -\frac{1}{3} \\ \hline \end{array}$ $\begin{array}{r} 3\frac{11}{12} \\ -\frac{6}{12} \\ \hline \end{array}$

4.
$$5\frac{7}{8}$$
$$-\ \ \frac{2}{8}$$
$$5\frac{5}{8}$$

$$6\frac{4}{6}$$
$$-\ \ \frac{1}{6}$$

$$2\frac{7}{10}$$
$$-\ \ \frac{4}{10}$$

$$6\frac{8}{9}$$
$$-\ \ \frac{2}{9}$$

$$9\frac{11}{12}$$
$$-\ \ \frac{10}{12}$$

Subtracting Mixed Numbers

To subtract mixed numbers, first subtract the fractions alone. Then subtract the whole numbers.

Reduce the proper fraction in the answer if possible.

Example:
$$4\frac{7}{9}$$
$$-2\frac{1}{9}$$
$$2\frac{6}{9}\ =\ 2\frac{2}{3}$$

5.
$$7\frac{2}{3}$$
$$-3\frac{1}{3}$$
$$4\frac{1}{3}$$

$$4\frac{4}{5}$$
$$-2\frac{1}{5}$$

$$3\frac{7}{8}$$
$$-1\frac{2}{8}$$

$$8\frac{11}{12}$$
$$-5\frac{4}{12}$$

$$2\frac{13}{16}$$
$$-1\frac{8}{16}$$

6.
$$5\frac{3}{4}$$
$$-1\frac{1}{4}$$
$$4\frac{2}{4}=4\frac{1}{2}$$

$$9\frac{7}{8}$$
$$-3\frac{3}{8}$$

$$4\frac{9}{10}$$
$$-2\frac{1}{10}$$

$$17\frac{5}{6}$$
$$-\ 9\frac{3}{6}$$

$$14\frac{7}{12}$$
$$-\ 6\frac{3}{12}$$

Subtracting Fractions from the Whole Number 1

To subtract a fraction from the whole number 1, change 1 to a fraction. For the numerator and denominator of this new fraction, use the number that is the denominator of the fraction that is being subtracted.

Example 1:
$$1\ =\ \frac{5}{5}$$
$$-\ \frac{3}{5}\ =\ \frac{3}{5}$$
$$\frac{2}{5}$$

Example 2:
$$1\ =\ \frac{8}{8}$$
$$-\ \frac{3}{8}\ =\ \frac{3}{8}$$
$$\frac{5}{8}$$

In Example 1, the whole number 1 is changed to $\frac{5}{5}$ since the denominator of the fraction being subtracted is 5.

In Example 2, the whole number 1 is changed to $\frac{8}{8}$. Do you see why?

7.
$$1\ =\ \frac{3}{3}$$
$$-\ \frac{2}{3}\ \ \frac{2}{3}$$
$$\frac{1}{3}$$

$$1$$
$$-\ \frac{3}{4}$$

$$1$$
$$-\ \frac{1}{5}$$

$$1$$
$$-\ \frac{4}{7}$$

$$1$$
$$-\ \frac{8}{9}$$

Subtracting Fractions from Whole Numbers

To subtract a fraction from a whole number, you must **borrow** 1 from the whole number. You change the borrowed 1 to a fraction and then subtract.

Example:

	Problem	*Step 1*	*Step 2*	*Step 3*
	6	$5 + 1$	$5\frac{8}{8}$	$5\frac{8}{8}$
	$-\frac{3}{8}$	$-\quad\frac{3}{8}$	$-\frac{3}{8}$	$-\frac{3}{8}$
				$5\frac{5}{8}$

Step 1. To subtract $\frac{3}{8}$, you need a fraction to subtract from. As a first step, borrow 1 from the 6.

Step 2. Change the borrowed 1 to the fraction $\frac{8}{8}$. You choose $\frac{8}{8}$ because $\frac{3}{8}$ has 8 as a denominator. Write $5 + \frac{8}{8}$ as $5\frac{8}{8}$.

Step 3. Subtract the fractions: $\frac{8}{8} - \frac{3}{8} = \frac{5}{8}$. Bring down the whole number 5.

Answer: $5\frac{5}{8}$

In row 1, show how each whole number can be rewritten by borrowing a 1 and changing it to a fraction. Do this by writing the correct numerator on each blank line.

1. $4 = 3\frac{}{5}$ $\qquad$ $7 = 6\frac{}{3}$ $\qquad$ $2 = 1\frac{}{9}$ $\qquad$ $4 = 3\frac{}{5}$ $\qquad$ $8 = 7\frac{}{4}$

In rows 2, 3, and 4, subtract. Reduce fraction answers when possible.

Skill Builders

2.

3	$=$	$2\frac{}{4}$		5	$=$	$4\frac{}{6}$		2	$=$	$1\frac{}{5}$		6	$=$	$5\frac{}{2}$
$-\frac{3}{4}$	$=$	$-\frac{3}{4}$		$-\frac{1}{6}$	$=$	$-\frac{1}{6}$		$-\frac{4}{5}$	$=$	$-\frac{4}{5}$		$-\frac{1}{2}$	$=$	$-\frac{1}{2}$

3.

5	8	2	3	2
$-\frac{4}{5}$	$-\frac{2}{3}$	$-\frac{3}{4}$	$-\frac{4}{7}$	$-\frac{11}{12}$

4.

6	12	7	9	10
$-\frac{1}{2}$	$-\frac{7}{8}$	$-\frac{3}{5}$	$-\frac{4}{6}$	$-\frac{15}{16}$

Subtracting Mixed Numbers by Borrowing

When the fraction in the bottom number is larger than the fraction above it, you must borrow in order to subtract. The borrowed 1 is changed to a fraction and added to the fraction that is already part of the top number.

Example:

Problem	Step 1	Step 2	Step 3
$4\frac{1}{5}$	$\overset{3}{\cancel{4}}\frac{5}{5} + \frac{1}{5}$	$3\frac{6}{5}$	$3\frac{6}{5}$
$-2\frac{4}{5}$	$-2\frac{4}{5}$	$-2\frac{4}{5}$	$-2\frac{4}{5}$
			$1\frac{2}{5}$

Step 1. You can't subtract $\frac{4}{5}$ from $\frac{1}{5}$. So, borrow 1 from the 4. Change the 4 to 3 and write the borrowed 1 as $\frac{5}{5}$. You choose $\frac{5}{5}$ because the fraction $\frac{1}{5}$ has 5 for a denominator. $4 = 3\frac{5}{5}$

Step 2. Combine the top fractions: $\frac{5}{5} + \frac{1}{5} = \frac{6}{5}$

Step 3. Subtract the fraction column: $\frac{6}{5} - \frac{4}{5} = \frac{2}{5}$

Subtract the whole numbers: $3 - 2 = 1$

Subtract. Complete the row of partially worked Skill Builders.

Skill Builders	$\overset{\frac{11}{8}}{\frown}$		$\overset{\frac{4}{3}}{\frown}$	
1. $5\frac{3}{8}$	$\rightarrow$	$\overset{4}{\cancel{5}}\frac{8}{8} + \frac{3}{8}$	$9\frac{1}{3}$ $\rightarrow$	$\overset{8}{\cancel{9}}\frac{3}{3} + \frac{1}{3}$
$-1\frac{5}{8}$		$-1\frac{5}{8}$	$-4\frac{2}{3}$	$-4\frac{2}{3}$

2.

$3\frac{1}{4}$	$4\frac{2}{5}$	$12\frac{1}{2}$	$14\frac{3}{8}$	$21\frac{3}{10}$
$-1\frac{3}{4}$	$-3\frac{4}{5}$	$-8\frac{1}{2}$	$-9\frac{7}{8}$	$-15\frac{7}{10}$

3.

$5\frac{1}{3}$	$15\frac{9}{12}$	$1\frac{1}{4}$	$2\frac{1}{8}$	$3\frac{2}{4}$
$-2\frac{2}{3}$	$-6\frac{11}{12}$	$-\frac{3}{4}$	$-1\frac{5}{8}$	$-\frac{3}{4}$

Solving Addition and Subtraction Word Problems

You solve fraction word problems in the same way you solve whole number and decimal word problems. Before thinking about what to do with the fractions, make sure you understand what the question asks you to find.

In each problem below, circle the words within the parentheses that best identify what you are asked to find. Then solve the problem.

1. Alice bought $2\frac{2}{3}$ yards of material to make a dress, $2\frac{1}{3}$ yards to make a wrap-around skirt, and $1\frac{2}{3}$ yards to make a blouse. How many total yards of material did Alice buy?

   (amount of material needed, (amount of material bought,) amount of material left over)

   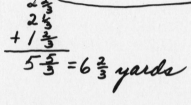

   $$2\frac{2}{3}$$
   $$2\frac{1}{3}$$
   $$+1\frac{2}{3}$$
   $$\overline{5\frac{5}{3}} = 6\frac{2}{3} \text{ yards}$$

2. A large pizza is cut into 12 equal slices. James ate 3, Stacey ate 2, and Frank ate 4. What fraction of pizza is left over?

   (amount of pizza ordered, amount of pizza eaten, amount of pizza not eaten)

3. Nails in bin C are $\frac{3}{8}$ inch longer than nails in bin B. If the nails in bin B are $\frac{5}{8}$ inch long, how long are the nails in bin C?

   (length of nails in bin B, length of nails in bin C, difference in length of the two nail sizes)

4. As shown at right, how much longer are the nails in bin A than the nails in bin B?

   (length of nails in bin A, length of nails in bin B, difference in length of the two nail sizes)

5. By comparing the drawings, figure out how much shorter Amy's doll is than Jane's doll.

   (height of Amy's doll, height of Jane's doll, difference in height of the two dolls)

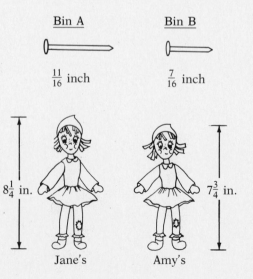

Bin A — $\frac{11}{16}$ inch

Bin B — $\frac{7}{16}$ inch

Jane's — $8\frac{1}{4}$ in.

Amy's — $7\frac{3}{4}$ in.

Problems 6 through 11 contain **extra information**. In each problem, write the **necessary information** on the blank line and then solve the problem.

6. Ben priced two gold rings. The plain ring cost $199 and contained $\frac{9}{32}$ ounces of gold. The fancy ring cost $289 and had $\frac{13}{32}$ ounces of gold. How much more gold did the fancy ring contain?

_____

7. When empty, Jean's suitcase weighs $5\frac{1}{8}$ pounds. Packing to go to Chicago, Jean packed $2\frac{3}{8}$ pounds of clothes, a $1\frac{7}{8}$-pound camera, and a pair of shoes that weighed $1\frac{1}{8}$ pounds. What total weight of stuff did Jean pack inside her suitcase?

_____

8. For her GED classes, Lillian studies very hard. Last week she studied $1\frac{3}{4}$ hours on Monday, 2 hours on Tuesday, $2\frac{3}{4}$ hours on Wednesday, and 4 hours on Friday. How many total hours did Lillian study during the first 3 days of last week?

_____

9. Following the recipe, Saul added $2\frac{1}{3}$ cups milk, $\frac{1}{3}$ cup cream, and $1\frac{2}{3}$ cups sugar in a bowl that holds $7\frac{1}{3}$ cups. How much liquid did he put in that bowl?

_____

10. As shown on the map at the right, Jenny lives farther from school than her friend David. Using the distances shown, figure out how far David lives from Jenny.

_____

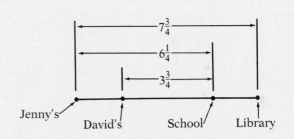

All distances in miles.

11. Wally's Seafoods is having a special sale on salmon steaks. His list of prices is shown at right. As indicated on the list, how much more salmon do you get for $8.05 than you get for $5.85?

_____

Salmon Prices	Package Size
$5.85	$1\frac{3}{4}$ pounds
$7.05	$2\frac{2}{4}$ pounds
$8.05	$3\frac{1}{4}$ pounds

Adding and Subtracting Unlike Fractions

Up to this point, you have worked with **like fractions** — fractions that have the same denominator. Like fractions can be added or subtracted.

Unlike fractions — fractions that have different denominators — can be added or subtracted only after they are changed to like fractions.

To change unlike fractions to like fractions, we write them as fractions that have the same number as a denominator. This number is called a **common denominator**. To do this, we raise at least one fraction to higher terms. (For a review of how to raise a fraction to higher terms, reread page 70 at this time.)

Example 1: Add $\frac{2}{3}$ and $\frac{1}{6}$.

Step 1. Choose 6 as a common denominator, and rewrite $\frac{2}{3}$ as sixths. To do this, write 6 as the new denominator, and then solve for the unknown numerator.

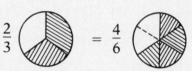

$$\frac{2}{3} = \frac{?}{6} \qquad \frac{2}{3} = \frac{4}{6}$$

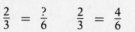

You can use 6 as a common denominator because one denominator is already 6, and the other denominator (3) divides into 6 an even number of times.

Step 2. Add the like fractions:

$$\frac{2}{3} = \frac{4}{6}$$
$$+\frac{1}{6} = \frac{1}{6}$$
$$\frac{5}{6}$$

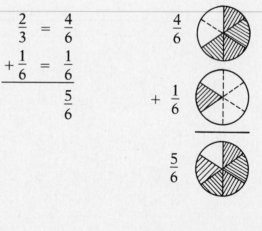

Answer: $\frac{5}{6}$

Example 2: Subtract $\frac{1}{2}$ from $\frac{7}{8}$.

Step 1. Choose 8 as a common denominator, and rewrite $\frac{1}{2}$ as eighths. Write 8 as the new denominator, and then solve for the unknown numerator.

$$\frac{1}{2} = \frac{?}{8} \qquad \frac{1}{2} = \frac{4}{8}$$

Step 2. Subtract $\frac{4}{8}$ from $\frac{7}{8}$.

$$\begin{array}{r} \frac{7}{8} = \frac{7}{8} \\ -\,\frac{1}{2} = \frac{4}{8} \\ \hline \frac{3}{8} \end{array}$$

$\dfrac{7}{8}$

$-\,\dfrac{4}{8}$

$\dfrac{3}{8}$

Answer: $\frac{3}{8}$

In rows 1 and 2, raise each fraction to higher terms using the denominator given. Do this by writing the correct numerator on each blank line.

1. $\frac{1}{3} = \frac{}{6}$ $\qquad$ $\frac{1}{2} = \frac{}{8}$ $\qquad$ $\frac{2}{3} = \frac{}{9}$ $\qquad$ $\frac{3}{4} = \frac{}{8}$ $\qquad$ $\frac{2}{3} = \frac{}{12}$

2. $\frac{3}{5} = \frac{}{10}$ $\qquad$ $\frac{3}{4} = \frac{}{12}$ $\qquad$ $\frac{4}{5} = \frac{}{15}$ $\qquad$ $\frac{7}{8} = \frac{}{16}$ $\qquad$ $\frac{3}{7} = \frac{}{28}$

In rows 3 through 6, add or subtract as indicated. Use the largest denominator in each problem as a common denominator. Reduce answers.

Skill Builders

3.
$\begin{array}{r} \frac{1}{3} = \frac{2}{6} \\ +\,\frac{3}{6} = \frac{}{6} \\ \hline \end{array}$
$\qquad$
$\begin{array}{r} \frac{3}{5} = \frac{}{10} \\ -\,\frac{3}{10} = \frac{3}{10} \\ \hline \end{array}$
$\qquad$
$\begin{array}{r} \frac{7}{8} = \frac{}{8} \\ -\,\frac{3}{4} = \frac{}{8} \\ \hline \end{array}$
$\qquad$
$\begin{array}{r} \frac{1}{2} = \frac{}{8} \\ \frac{1}{4} = \frac{}{8} \\ +\,\frac{3}{8} = \frac{}{8} \\ \hline \end{array}$
$\qquad$
$\begin{array}{r} \frac{11}{12} = \frac{}{12} \\ \frac{5}{6} = \frac{}{12} \\ +\,\frac{1}{4} = \frac{}{12} \\ \hline \end{array}$

4.
$\begin{array}{r} \frac{1}{2} \\ +\,\frac{1}{6} \\ \hline \end{array}$
$\qquad$
$\begin{array}{r} \frac{4}{5} \\ -\,\frac{3}{10} \\ \hline \end{array}$
$\qquad$
$\begin{array}{r} \frac{3}{4} \\ +\,\frac{1}{2} \\ \hline \end{array}$
$\qquad$
$\begin{array}{r} \frac{5}{6} \\ -\,\frac{2}{3} \\ \hline \end{array}$
$\qquad$
$\begin{array}{r} \frac{7}{8} \\ -\,\frac{3}{4} \\ \hline \end{array}$

5. $\frac{4}{7} + \frac{3}{14}$ $\qquad$ $\frac{7}{9} - \frac{2}{3}$ $\qquad$ $\frac{11}{12} + \frac{1}{4}$ $\qquad$ $\frac{15}{16} - \frac{9}{32}$ $\qquad$ $\frac{7}{12} + \frac{5}{6}$

6.
$\begin{array}{r} \frac{2}{6} \\ \frac{1}{3} \\ +\,\frac{1}{6} \\ \hline \end{array}$
$\qquad$
$\begin{array}{r} \frac{5}{8} \\ \frac{1}{4} \\ +\,\frac{1}{8} \\ \hline \end{array}$
$\qquad$
$\begin{array}{r} \frac{2}{5} \\ \frac{1}{5} \\ +\,\frac{3}{10} \\ \hline \end{array}$
$\qquad$
$\begin{array}{r} \frac{1}{2} \\ \frac{3}{4} \\ +\,\frac{1}{8} \\ \hline \end{array}$
$\qquad$
$\begin{array}{r} \frac{7}{12} \\ \frac{1}{3} \\ +\,\frac{3}{4} \\ \hline \end{array}$

Choosing a Common Denominator

Often, the largest denominator in a problem cannot be a common denominator. **A common denominator must be a number that each denominator divides into evenly**.

Multiplying Denominators Gives a Common Denominator

One way to choose a common denominator is to multiply the denominators in a problem by each other.

For example, at right $4 \times 3 = 12$.

12 can be used as a common denominator because both 4 and 3 divide into 12 evenly.

Multiplying denominators is the best method to use to find a common denominator in most problems:

- It works well when denominators are small numbers.

- It works well when there are more than two denominators in a problem.

In the example, 12 is the smallest whole number that can be used as a common denominator. Because of this, 12 is called the **lowest common denominator (LCD)**.

In each problem below, **multiply denominators** to find a common denominator, and then add or subtract. The Skill Builders are partially worked for you. Reduce answers.

Skill Builders

1.

$$\frac{1}{2} = \frac{}{6}$$
$$-\frac{1}{3} = \frac{}{6}$$

$$\frac{2}{3} = \frac{}{12}$$
$$+\frac{1}{4} = \frac{}{12}$$

$$\frac{4}{5} = \frac{}{20}$$
$$-\frac{3}{4} = \frac{}{20}$$

$$\frac{1}{2} = \frac{}{30}$$
$$\frac{1}{3} = \frac{}{30}$$
$$+\frac{1}{5} = \frac{}{30}$$

$$\frac{3}{4} = \frac{}{24}$$
$$\frac{2}{3} = \frac{}{24}$$
$$+\frac{1}{2} = \frac{}{24}$$

2.

$$\frac{2}{3}$$
$$+\frac{1}{2}$$

$$\frac{4}{5}$$
$$-\frac{1}{2}$$

$$\frac{3}{4}$$
$$+\frac{1}{3}$$

$$\frac{7}{9}$$
$$-\frac{2}{5}$$

$$\frac{3}{8}$$
$$-\frac{1}{3}$$

3.

$$\begin{array}{r}\frac{2}{3}\\\frac{1}{2}\\+\frac{2}{5}\end{array}\qquad\begin{array}{r}\frac{3}{4}\\\frac{1}{2}\\+\frac{1}{6}\end{array}\qquad\begin{array}{r}\frac{1}{2}\\\frac{1}{3}\\+\frac{1}{4}\end{array}\qquad\begin{array}{r}\frac{1}{3}\\\frac{1}{4}\\+\frac{1}{5}\end{array}\qquad\begin{array}{r}\frac{3}{4}\\\frac{2}{5}\\+\frac{1}{6}\end{array}$$

Finding the Lowest Common Denominator

Multiplying denominators always gives a common denominator. However, this method does not always give the **lowest common denominator**.

In the example at right, multiplying 8 times 6 gives 48 as a common denominator.

Notice that 24 will also work as a common denominator. Using 24, you do not need to reduce the answer, and you work with smaller fractions. At right, 24 is the **lowest common denominator**.

Using 48

$$\frac{5}{8}=\frac{30}{48}$$
$$-\frac{1}{6}=\frac{8}{48}$$
$$\overline{\frac{22}{48}}=\frac{11}{24}$$

Using 24

$$\frac{5}{8}=\frac{15}{24}$$
$$-\frac{1}{6}=\frac{4}{24}$$
$$\overline{\frac{11}{24}}$$

To find the lowest common denominator in any problem, compare the smallest denominator with multiples of the largest denominator. A multiple is found by multiplying a denominator by 1, 2, 3, and so on.

> **The lowest common denominator is the smallest multiple that each denominator divides into evenly.**

Look again at the example:

$\frac{5}{8}$ The multiples of 8 are: 8, 16, 24, 32 . . .

$-\frac{1}{6}$ Find a multiple of 8 that 6 divides evenly into: 6

The smallest multiple of 8 that 6 divides into evenly is 24. Thus, 24 is the lowest common denominator for the example.

Find a common denominator and solve the problem. In the Skill Builders, some multiples of the largest denominator are written for you.

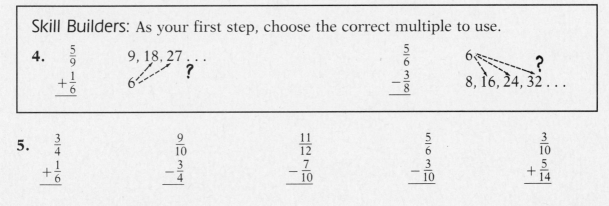

Skill Builders: As your first step, choose the correct multiple to use.

4.

$$\begin{array}{r}\frac{5}{9}\\+\frac{1}{6}\end{array}\qquad 9, 18, 27 \ldots$$
$$6 \,\text{--→?}$$

$$\begin{array}{r}\frac{5}{6}\\-\frac{3}{8}\end{array}\qquad 6\text{--→?}$$
$$8, 16, 24, 32 \ldots$$

5.

$$\begin{array}{r}\frac{3}{4}\\+\frac{1}{6}\end{array}\qquad\begin{array}{r}\frac{9}{10}\\-\frac{3}{4}\end{array}\qquad\begin{array}{r}\frac{11}{12}\\-\frac{7}{10}\end{array}\qquad\begin{array}{r}\frac{5}{6}\\-\frac{3}{10}\end{array}\qquad\begin{array}{r}\frac{3}{10}\\+\frac{5}{14}\end{array}$$

85

Adding and Subtracting Mixed Numbers

You can add or subtract mixed numbers only if the fractions are **like fractions**. If they are **unlike fractions**, choose a common denominator and rewrite them as like fractions. Use either the method of **multiplying denominators** or the method of **comparing multiples** to choose a common denominator.

- When adding, remember to change any improper fraction answer to a mixed number. Then add the whole numbers together in the answer.
- When subtracting, rewrite unlike fractions as like fractions *before* you do any needed borrowing.
- For all problems, remember to work with fractions first, and then work with whole numbers. Simplify answers when possible.

Adding Mixed Numbers

Add. Complete the row of partially worked Skill Builders.

Skill Builders

1.

$3\frac{2}{3} = 3\frac{}{6}$ $5\frac{3}{4} = 5\frac{}{12}$ $4\frac{3}{5} = 4\frac{}{20}$ $6\frac{1}{3} = 6\frac{}{12}$

$+1\frac{1}{2} = 1\frac{}{6}$ $+2\frac{1}{3} = 2\frac{}{12}$ $+3\frac{1}{4} = 3\frac{}{20}$ $2\frac{5}{6} = 2\frac{}{12}$

$+1\frac{3}{4} = 1\frac{}{12}$

2.

$2\frac{1}{2}$ $4\frac{2}{3}$ $7\frac{3}{5}$ $6\frac{3}{4}$ $5\frac{3}{7}$

$+1\frac{3}{4}$ $+2\frac{1}{6}$ $+3\frac{1}{2}$ $+5\frac{1}{3}$ $+4\frac{1}{2}$

3.

$3\frac{4}{5}$ $8\frac{3}{8}$ $4\frac{5}{6}$ $12\frac{2}{3}$ $20\frac{3}{7}$

$+2\frac{3}{4}$ $+3\frac{2}{5}$ $+2\frac{1}{4}$ $+9\frac{3}{5}$ $+6\frac{1}{3}$

4.

$4\frac{1}{2}$ $5\frac{3}{8}$ $7\frac{1}{3}$ $12\frac{1}{4}$ $21\frac{2}{3}$

$2\frac{2}{3}$ $3\frac{1}{4}$ $5\frac{2}{5}$ $11\frac{1}{2}$ $15\frac{3}{5}$

$+1\frac{3}{4}$ $+4\frac{1}{2}$ $+2\frac{3}{4}$ $+9\frac{5}{6}$ $+8\frac{1}{2}$

Subtracting Mixed Numbers

Subtract. Complete the row of partially worked Skill Builders.

Skill Builders

5.

$$7\frac{2}{3} = 7\frac{}{12}$$
$$-5\frac{1}{4} = 5\frac{}{12}$$

$$13\frac{5}{6} = 13\frac{}{12}$$
$$-8\frac{3}{4} = 8\frac{}{12}$$

$$9\frac{3}{4} = 9\frac{15}{20} = 9\frac{8}{20}$$
$$-6\frac{4}{5} = 6\frac{16}{20} = 6\frac{}{20}$$

6.

$$9\frac{3}{4}$$
$$-6\frac{2}{3}$$

$$12\frac{4}{9}$$
$$-8\frac{1}{3}$$

$$17\frac{4}{5}$$
$$-10\frac{1}{4}$$

$$7\frac{5}{6}$$
$$-\frac{1}{4}$$

$$36\frac{1}{2}$$
$$-23\frac{2}{5}$$

7.

$$11\frac{1}{2}$$
$$-7\frac{3}{4}$$

$$6\frac{3}{8}$$
$$-5\frac{7}{16}$$

$$2\frac{1}{4}$$
$$-\frac{2}{5}$$

$$15\frac{5}{6}$$
$$-8\frac{7}{8}$$

$$29\frac{1}{2}$$
$$-17\frac{2}{3}$$

Mixed Practice

8.

$$\frac{2}{3}$$
$$+\frac{3}{4}$$

$$1\frac{7}{8}$$
$$-\frac{2}{3}$$

$$2\frac{3}{5}$$
$$-\frac{9}{10}$$

$$3\frac{15}{16}$$
$$+\frac{7}{8}$$

$$2\frac{3}{4}$$
$$-\frac{7}{8}$$

9.

$$\frac{11}{16}$$
$$-\frac{1}{4}$$

$$2\frac{3}{8}$$
$$+1\frac{5}{8}$$

$$5\frac{3}{7}$$
$$-2\frac{5}{14}$$

$$21\frac{11}{12}$$
$$+16\frac{5}{8}$$

$$\frac{9}{10}$$
$$+\frac{3}{4}$$

10.

$$1\frac{2}{3}$$
$$-\frac{3}{4}$$

$$4\frac{3}{16}$$
$$-2\frac{5}{8}$$

$$14\frac{5}{9}$$
$$+6\frac{2}{3}$$

$$8\frac{3}{10}$$
$$-5\frac{3}{5}$$

$$13\frac{7}{12}$$
$$+7\frac{7}{8}$$

Using Substitution to Help Solve Word Problems

Do fraction word problems seem more difficult than other problems? If you answer "Yes," you're like most other people. One reason fractions are difficult is that they are hard to visualize (to see in your mind). Because of this, your math intuition probably isn't as good with fractions as it is with whole numbers.

On these next two pages, we'll show you a technique called **substitution** that may help you solve fraction word problems. Here's how you use substitution:

- Replace fractions in a problem with small whole numbers.

- Next, read the problem as a whole number problem and decide what problem-solving steps to use—for example, whether to add or to subtract.

- Finally, go back and carry out the same problem-solving steps on the fractions in the original problem.

Example:

Original Problem Written with Fractions

A bolt that measures $2\frac{5}{16}$ inches long is to be shortened by $\frac{3}{8}$ inch. What length should the bolt be cut?

Substitution Problem Written with Small Whole Numbers

A bolt that measures 2 inches long is to be shortened by 1 inch. What length should the bolt be cut?

As written at right, you can see that this is a subtraction problem. You find the answer by subtracting the smaller number 1 from the larger number 2.

Substitution Problem Solution

$2 - 1 = 1$ inch
Answer: 1 inch

Using the same subtraction step in the original problem, you find the answer by subtracting $\frac{3}{8}$ from $2\frac{5}{16}$.

Answer: $1\frac{15}{16}$ inches

Original Problem Solution

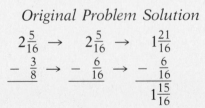

When you use substitution, it does not matter what small whole numbers you use. Because it seems easiest to most students, in the problems that follow we'll always substitute 1 for the smallest fraction or mixed number and 2 or 3 for the largest.

First solve each addition or subtraction problem below by substituting whole numbers as indicated. Write the answer to the substitution problem on the first line, and write the answer to the original problem on the second line.

1. Jessica grew $1\frac{1}{4}$ inches last year. This was $\frac{5}{16}$ of an inch more than she grew the previous year. Figure out how much Jessica grew during that previous year.

   Substitute 2 for $1\frac{1}{4}$ and 1 for $\frac{5}{16}$.

   $\underline{2-1=1}$ $\underline{1\frac{1}{4} - \frac{5}{16} = \frac{15}{16}}$
      substitute   real

2. To make a large fruit salad, Janet bought $1\frac{1}{2}$ pounds of grapes, $1\frac{2}{3}$ pounds of strawberries, and $2\frac{1}{4}$ pounds of apples. Mixed together, how many pounds of salad will this fruit make?

   Substitute 1 for $1\frac{1}{2}$, 2 for $1\frac{2}{3}$, and 3 for $2\frac{1}{4}$.

   _____   _____
      substitute   real

3. The value of Irving Company stock went from $19\frac{1}{2}$ to $18\frac{5}{8}$ between July 1 and July 15. How many points in value did the stock change during this two-week period?

   Substitute 2 for $19\frac{1}{2}$ and 1 for $18\frac{5}{8}$.

   _____   _____
      substitute   real

4. For a birthday present, Ernie bought $1\frac{1}{2}$ pounds of selected chocolates. Because of a sale, he was given an extra $\frac{3}{16}$ pound at no extra cost. How much chocolate did Ernie leave the store with, assuming that he didn't eat any?

   Substitute 2 for $1\frac{1}{2}$ and 1 for $\frac{3}{16}$.

   _____   _____
      substitute   real

When you're not sure which of two fractions is larger, rewrite the fractions to have a common denominator. Then use substitution as before. In problems 5, 6, and 7 substitute 2 for the larger fraction and 1 for the smaller fraction.

5. Connie ordered carpet that was $\frac{11}{16}$ inch thick. The pad underneath it was $\frac{5}{8}$ inch thick. By how much will the carpet and pad raise the level of the floor?

   _____   _____
      substitute   real

6. On Saturday it rained $\frac{5}{6}$ of an inch. On Sunday it rained another $\frac{7}{8}$ inch. How much more rain fell on one day than on the other?

   _____   _____
      substitute   real

7. According to the blueprint, the spring should be no longer than $\frac{9}{32}$ inch. But, when he measured it, Lars found it to be $\frac{5}{16}$ inch. By how much does the spring differ from its correct length?

   _____   _____
      substitute   real

Solving Addition and Subtraction Word Problems

On the next two pages are groups of addition and subtraction word problems. Each group is designed to focus your attention on a special problem-solving skill. And, in many problems, you may find it helpful to use **substitution** to help you visualize what is being described.

As you work these problems, remember to change **unlike fractions** to **like fractions** as your first step before either adding or subtracting.

Below are three pairs of problems that contain the same necessary information. Yet each pair contains an addition problem and a subtraction problem. Solve these problems by paying close attention to what each question asks you to find.

1. a) Kate ran $2\frac{3}{4}$ miles on Monday. On Wednesday she ran $3\frac{1}{8}$ miles. How much farther did Kate run on Wednesday than on Monday?

b) Kate ran $2\frac{3}{4}$ miles on Monday. On Wednesday she ran $3\frac{1}{8}$ miles. How many miles did she run on those two days?

2. a) While shopping at Frank's Meats, Sharon bought $3\frac{1}{2}$ pounds of steak and $2\frac{3}{4}$ pounds of pork. How much more steak did she buy than pork?

b) While shopping at Frank's Meats, Sharon bought $3\frac{1}{2}$ pounds of steak and $2\frac{3}{4}$ pounds of pork. How much meat did Sharon buy at Frank's?

3. a) James has two metal rods. One measures $9\frac{5}{16}$ inches and the other measures $5\frac{7}{8}$ inches. What is the length of the two rods placed end to end?

b) James has two metal rods. One measures $9\frac{5}{16}$ inches and the other measures $5\frac{7}{8}$ inches. What is the difference in length between the two rods?

Complete problems 4 and 5 by choosing the word within parentheses that makes each an **addition problem**. Then solve each problem.

4. Georgia went to the store to buy about $6\frac{1}{2}$ pounds of potatoes. Instead, she found a bargain bag of potatoes that weighed $1\frac{2}{3}$ pounds _____. How many pounds of potatoes did she end up buying?

(less, more)

5. During May, Friendly Computer stock sold at $34\frac{1}{2}$ per share. By June 15, the price _____ by $1\frac{7}{8}$ points. What price did the stock sell for June 15?

(rose, fell)

Complete problems 6 and 7 by choosing the word within parentheses that makes each a **subtraction problem**. Then solve each problem.

6. Norma usually works $4\frac{1}{2}$ hours each Saturday morning. For this Saturday, though, her boss told her that she would go home $\frac{3}{4}$ hour _____. How many hours can Norma expect to work this Saturday?

(earlier, later)

7. Last month, Bert lost $3\frac{7}{8}$ pounds on his diet. This month he plans to lose $1\frac{1}{4}$ _____ pounds than he lost last month. How many total pounds is Bert trying to lose this month?

(fewer, more)

In problems 8 and 9, underline the necessary information and solve each problem.

8. In the last budget election, $\frac{11}{12}$ of the registered voters voted. Only $\frac{1}{3}$ of those who voted were *for* the proposed budget, while $\frac{7}{12}$ were *against* it. How many more voters voted *against* the budget than voted *for* it? Your answer will be a fraction of the voters.

9. The loaded backpack weighs $31\frac{3}{4}$ pounds. Inside the pack is a $2\frac{1}{2}$ pound sleeping bag, other camping gear, and $14\frac{1}{3}$ pounds of food supplies. If the backpack itself weighs $4\frac{7}{8}$ pounds, what is the total weight of the stuff packed inside?

In problems 10 and 11, circle the arithmetic expression that will give the correct answer to each question. You do not need to solve these two problems.

10. Scott has a roll of screen material to make some household repairs. If he cuts off two pieces each $1\frac{1}{2}$ yards wide for windows, and one $3\frac{1}{3}$-yard-wide piece for a door, what total length of screen material will he remove from the roll?

a) $\frac{5}{2} + \frac{3}{2} - \frac{10}{3}$

b) $2 + \frac{3}{2} - \frac{10}{3}$

c) $\frac{3}{2} + \frac{3}{2} + \frac{10}{3}$

d) $\frac{3}{2} + \frac{3}{2} - \frac{10}{3}$

11. The cookie recipe calls for $1\frac{3}{4}$ cups of flour. If Leonard has only $1\frac{1}{8}$ cups of flour, how much more flour does he need?

a) $\frac{14}{8} - \frac{9}{8}$

b) $\frac{13}{8} - \frac{9}{8}$

c) $\frac{7}{4} - \frac{1}{4}$

d) $\frac{7}{4} - (\frac{9}{8} + \frac{1}{4})$

Multiplying Fractions

To multiply by a fraction is to take a part of something. For example, to multiply $\frac{1}{2}$ by $\frac{1}{4}$ is to take $\frac{1}{2}$ of $\frac{1}{4}$.

As shown at right, $\frac{1}{2}$ of $\frac{1}{4}$ is $\frac{1}{8}$.

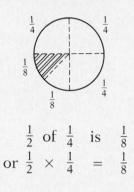

$\frac{1}{2}$ of $\frac{1}{4}$ is $\frac{1}{8}$

or $\frac{1}{2} \times \frac{1}{4} = \frac{1}{8}$

The rule for multiplying fractions is easy to learn.

- Multiply the numerators of the fractions to find the numerator of the answer.

- Multiply the denominators of the fractions to find the denominator of the answer.

An especially easy way to remember this rule is to say the following short version to yourself several times:

"Top times top equals top."

"Bottom times bottom equals bottom."

Example 1: Multiply $\frac{1}{2}$ by $\frac{3}{4}$.

Step 1. Multiply the top numbers to find the numerator of the answer.

$$\frac{1}{2} \times \frac{3}{4} = \frac{3}{}$$

Step 2. Multiply the bottom numbers to find the denominator of the answer.

$$\frac{1}{2} \times \frac{3}{4} = \frac{3}{8}$$

Answer: $\frac{3}{8}$

Example 2: $\frac{4}{5} \times \frac{3}{8}$

Step 1. Multiply.

$$\frac{4}{5} \times \frac{3}{8} = \frac{12}{40}$$

Step 2. Reduce the answer.

$$\frac{12 \div 4}{40 \div 4} = \frac{3}{10}$$

Answer: $\frac{3}{10}$

You use the same rule to multiply both proper and improper fractions. You also use this rule to multiply more than two fractions in the same problem.

Multiply. The first problem in each row is done for you.

1. $\frac{1}{2} \times \frac{1}{3} = \frac{1}{6}$ $\frac{2}{3} \times \frac{1}{5} =$ $\frac{3}{4} \times \frac{1}{3} =$ $\frac{2}{5} \times \frac{2}{9} =$ $\frac{3}{8} \times \frac{1}{2} =$

2. $\frac{2}{3} \times \frac{1}{7} = \frac{2}{21}$ $\frac{1}{2} \times \frac{1}{2} =$ $\frac{2}{3} \times \frac{4}{5} =$ $\frac{3}{7} \times \frac{3}{10} =$ $\frac{5}{6} \times \frac{3}{4} =$

3. $\frac{1}{3} \times \frac{3}{4} \times \frac{2}{5} = \frac{6}{60}$ $\frac{2}{3} \times \frac{1}{2} \times \frac{3}{4} =$ $\frac{3}{2} \times \frac{2}{3} \times \frac{4}{5} =$

            $= \frac{1}{10}$

Using Canceling to Simplify Multiplication

You can use a shortcut called ***canceling*** when you multiply fractions.

To cancel, divide both a top number and a bottom number by the same number. Cancel before you multiply.

Canceling is similar to **reducing** a fraction. However, when you cancel, you simplify fractions **before** you multiply.

Example 1: $\frac{2}{9} \times \frac{6}{7}$

With Canceling

Step 1. Divide the right numerator (6) by 3 and divide the left denominator (9) by 3.

$$\frac{2}{_3\cancel{9}} \times \frac{\cancel{6}^2}{7}$$

Step 2. Multiply the rewritten fractions:

$$\frac{2}{3} \times \frac{2}{7} = \frac{4}{21}$$

Answer: $\frac{4}{21}$

Without Canceling

$$\frac{2}{9} \times \frac{6}{7} = \frac{12}{63}$$
$$= \frac{12 \div 3}{63 \div 3}$$
$$= \frac{4}{21}$$

In Example 2, notice how canceling may be used more than once in the same problem.

Example 2: $\frac{4}{9} \times \frac{3}{8}$

Step 1. Divide both the 4 and the 8 by 4. $\quad \frac{^1\cancel{4}}{_3\cancel{9}} \times \frac{\cancel{3}^1}{\cancel{8}_2} = \frac{1}{3} \times \frac{1}{2} = \frac{1}{6}$

Step 2. Divide both the 3 and the 9 by 3.

Use canceling to solve each problem below.

Skill Builders

4. $\quad \frac{5}{6} \times \frac{3}{4} \qquad\qquad \frac{4}{7} \times \frac{10}{12} \qquad\qquad \frac{12}{9} \times \frac{3}{4} \qquad\qquad \frac{7}{8} \times \frac{16}{14}$

$\quad = \frac{5}{_2\cancel{6}} \times \frac{\cancel{3}^1}{4} \qquad = \frac{^1\cancel{4}}{7} \times \frac{10}{\cancel{12}_3} \qquad = \frac{^3\cancel{12}}{_3\cancel{9}} \times \frac{\cancel{3}^1}{\cancel{4}_1} \qquad = \frac{^1\cancel{7}}{_1\cancel{8}} \times \frac{\cancel{16}^2}{\cancel{14}_2}$

---

5. $\frac{7}{8} \times \frac{5}{7} \qquad\qquad \frac{4}{5} \times \frac{3}{8} \qquad\qquad \frac{5}{9} \times \frac{4}{15} \qquad\qquad \frac{3}{4} \times \frac{12}{7}$

6. $\frac{7}{4} \times \frac{16}{14} \qquad\qquad \frac{3}{8} \times \frac{2}{9} \qquad\qquad \frac{12}{5} \times \frac{25}{36} \qquad\qquad \frac{2}{7} \times \frac{14}{18}$

Multiplying with Fractions and Whole Numbers

To multiply a fraction by a whole number, you must first place the whole number over the number 1. In this way **the whole number is also written as a fraction**. Then you multiply just as you multiply any two fractions. Say the rule to yourself so that you'll always remember how to multiply fractions:

Example:

$$\frac{3}{4} \times 5$$
$$= \frac{3}{4} \times \frac{5}{1}$$
$$= \frac{15}{4}$$
$$= 3\frac{3}{4}$$

"Top times top equals top."

"Bottom times bottom equals bottom."

Change any improper fraction answer to a mixed number.

Change each whole number in row 7 to a fraction. The first one is done as an example.

7. $4 = \dfrac{4}{1}$ $\qquad 9 =$ $\qquad 7 =$ $\qquad 8 =$ $\qquad 5 =$

Multiply. Complete the row of partially worked Skill Builders.

Skill Builders

8. $\dfrac{3}{9} \times 6$ $\qquad\qquad 4 \times \dfrac{5}{8}$ $\qquad\qquad \dfrac{12}{7} \times 4$ $\qquad\qquad 8 \times \dfrac{11}{12}$

$\quad = \dfrac{3}{9} \times \dfrac{6}{1}$ $\qquad = \dfrac{4}{1} \times \dfrac{5}{8}$ $\qquad = \dfrac{12}{7} \times \dfrac{4}{1}$ $\qquad = \dfrac{8}{1} \times \dfrac{11}{12}$

9. $\dfrac{4}{5} \times 6$ $\qquad \dfrac{7}{8} \times 7$ $\qquad \dfrac{3}{4} \times 12$ $\qquad \dfrac{7}{8} \times 4$ $\qquad \dfrac{15}{12} \times 8$

10. $6 \times \dfrac{4}{5}$ $\qquad 8 \times \dfrac{11}{12}$ $\qquad 5 \times \dfrac{3}{4}$ $\qquad 7 \times \dfrac{12}{7}$ $\qquad 3 \times \dfrac{11}{6}$

11. $\dfrac{14}{3} \times 2$ $\qquad 8 \times \dfrac{9}{6}$ $\qquad 3 \times \dfrac{11}{12}$ $\qquad 21 \times \dfrac{1}{2}$ $\qquad \dfrac{5}{6} \times \dfrac{16}{15}$

Multiplying with Mixed Numbers

To multiply one or more mixed numbers, the first step is to change all mixed numbers to improper fractions. Then multiply as before.

Example: Change $2\frac{3}{4}$ to an improper fraction.

Step 1.   Change the whole number to an improper fraction; multiply by the denominator of the fraction. Put the product over the denominator. $2 \times \frac{}{4} = \frac{8}{4}$

Step 2.   Add to the other fraction. $\frac{8}{4} + \frac{3}{4} = \frac{11}{4}$

Change each mixed number in row 12 to an improper fraction.

12. $4\frac{2}{3} =$ $3\frac{5}{8} =$ $2\frac{3}{5} =$ $7\frac{1}{2} =$ $5\frac{3}{4} =$

Skill Builders

13. $1\frac{2}{3} \times 3$ $2 \times 3\frac{3}{4}$ $\frac{5}{6} \times 2\frac{3}{8}$ $4\frac{1}{2} \times \frac{3}{5}$ $2\frac{1}{2} \times 3\frac{1}{2}$

    $= \frac{5}{3} \times \frac{3}{1}$ $= \frac{2}{1} \times \frac{15}{4}$ $= \frac{5}{6} \times \frac{19}{8}$ $= \frac{9}{2} \times \frac{3}{5}$ $= \frac{5}{2} \times \frac{7}{2}$

    $=$ $=$ $=$ $=$ $=$

14. $\frac{3}{4} \times 1\frac{1}{4}$ $\frac{2}{3} \times 3\frac{1}{2}$ $4\frac{1}{2} \times \frac{3}{8}$ $\frac{4}{5} \times 2\frac{5}{6}$ $3\frac{2}{7} \times 4\frac{1}{3}$

15. $4 \times 2\frac{2}{3}$ $6\frac{5}{8} \times 3$ $2\frac{3}{8} \times 4$ $10 \times 2\frac{3}{8}$ $6\frac{7}{10} \times 2$

16. $2\frac{3}{4} \times 3\frac{4}{5}$ $1\frac{3}{8} \times 5\frac{1}{2}$ $4\frac{1}{3} \times 2\frac{11}{12}$ $5\frac{2}{7} \times 4\frac{5}{9}$ $3\frac{3}{10} \times 1\frac{9}{10}$

Dividing Fractions

When you divide, you are finding out how many times one number will go into a second number. This is true for fractions as well as whole numbers.

For example, suppose you want to know how many $\frac{1}{16}$-mile-long sections are in a stretch of road that measures $\frac{3}{8}$ mile long. To find out, you divide $\frac{3}{8}$ by $\frac{1}{16}$.

As shown at right, there are **six** $\frac{1}{16}$s in $\frac{3}{8}$.

$$\frac{3}{8}$$

$\frac{1}{8}$	$\frac{1}{8}$	$\frac{1}{8}$

| $\frac{1}{16}$ | $\frac{1}{16}$ | $\frac{1}{16}$ | $\frac{1}{16}$ | $\frac{1}{16}$ | $\frac{1}{16}$ |

Dividing fractions is not difficult. In fact, after mastering multiplication, you need to learn only one more step in order to divide fractions.

To divide fractions, invert the divisor (the number you're dividing by) **and change the division sign to a multiplication sign. Then multiply to find the answer.**

Inverting the Divisor

To **invert** means to "turn a fraction upside down." You do this by interchanging the numerator with the denominator. The top number becomes the bottom number, and the bottom number becomes the top number.

Inverting $\frac{2}{3}$

$$\frac{2}{3} \diagdown \diagup \frac{3}{2}$$

Before inverting a divisor, you must first change it to a fraction if it's not one already.

- Change a whole number divisor to a whole number over 1.

- Change a mixed number divisor to an improper fraction.

Type of Divisor	Example	Written as a Fraction	Inverted
Proper fraction	$\frac{3}{4}$	$\frac{3}{4}$	$\frac{4}{3}$
Improper fraction	$\frac{12}{7}$	$\frac{12}{7}$	$\frac{7}{12}$
Whole number	5	$\frac{5}{1}$	$\frac{1}{5}$
Mixed number	$2\frac{3}{8}$	$\frac{19}{8}$	$\frac{8}{19}$

Invert each number below. As a first step, change any whole number or mixed number to a fraction.

1. $\frac{4}{7}$ $\qquad$ $\frac{8}{3}$ $\qquad$ 9 $\qquad$ $\frac{11}{2}$ $\qquad$ $4\frac{2}{3}$

2. 9 $\qquad$ $3\frac{3}{4}$ $\qquad$ $\frac{4}{5}$ $\qquad$ $1\frac{7}{8}$ $\qquad$ 7

Dividing Fractions by Fractions

To divide one fraction by another, follow these steps:

1. Invert the fraction to the right of the division sign.

2. Change the division sign to a multi-plication sign and multiply.

Simplify the answer if possible.

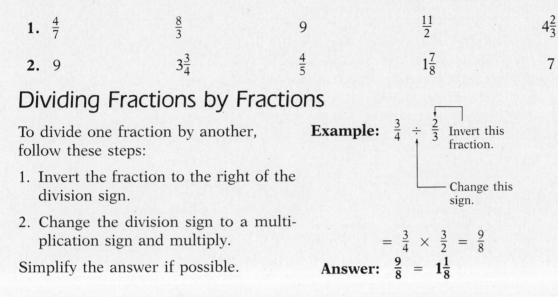

Example: $\frac{3}{4} \div \frac{2}{3}$ Invert this fraction.

Change this sign.

$= \frac{3}{4} \times \frac{3}{2} = \frac{9}{8}$

Answer: $\frac{9}{8} = 1\frac{1}{8}$

As you'll see in the problems below, when you divide a fraction by a fraction, the answer may be **less than 1, equal to 1,** or **more than 1.**

Skill Builders

3. $\frac{1}{2} \div \frac{3}{4} \rightarrow \frac{1}{2} \times \frac{4}{3}$ $\qquad$ $\frac{7}{8} \div \frac{7}{8} \rightarrow \frac{7}{8} \times \frac{8}{7}$ $\qquad$ $\frac{4}{5} \div \frac{1}{3} \rightarrow \frac{4}{5} \times \frac{3}{1}$

$=$ $\qquad\qquad\qquad$ $=$ $\qquad\qquad\qquad$ $=$

---

4. $\frac{2}{3} \div \frac{1}{2}$ $\qquad$ $\frac{4}{5} \div \frac{2}{3}$ $\qquad$ $\frac{7}{8} \div \frac{1}{3}$ $\qquad$ $\frac{1}{3} \div \frac{3}{8}$ $\qquad$ $\frac{3}{4} \div \frac{1}{3}$

5. $\frac{4}{5} \div \frac{4}{5}$ $\qquad$ $\frac{5}{6} \div \frac{8}{7}$ $\qquad$ $\frac{6}{5} \div \frac{3}{4}$ $\qquad$ $\frac{9}{8} \div \frac{4}{3}$ $\qquad$ $\frac{1}{2} \div \frac{4}{5}$

6. $\frac{12}{8} \div \frac{1}{2}$ $\qquad$ $\frac{7}{5} \div \frac{7}{5}$ $\qquad$ $\frac{2}{5} \div \frac{5}{2}$ $\qquad$ $\frac{13}{4} \div \frac{2}{3}$ $\qquad$ $\frac{3}{6} \div \frac{1}{3}$

7. $\frac{5}{9} \div \frac{5}{9}$ $\qquad$ $\frac{12}{3} \div \frac{2}{4}$ $\qquad$ $\frac{3}{4} \div \frac{13}{2}$ $\qquad$ $\frac{2}{7} \div \frac{14}{5}$ $\qquad$ $\frac{15}{4} \div \frac{12}{5}$

Dividing with Fractions and Whole Numbers

Often you may want to divide a fraction by a whole number. To do this, write the whole number as a fraction with a denominator of 1. Then invert the whole number divisor and multiply.

Example:

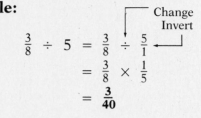

$$\frac{3}{8} \div 5 = \frac{3}{8} \div \frac{5}{1}$$

$$= \frac{3}{8} \times \frac{1}{5}$$

$$= \frac{3}{40}$$

Dividing a fraction by a whole number gives you an answer that is always smaller than the fraction you started with.

Skill Builders

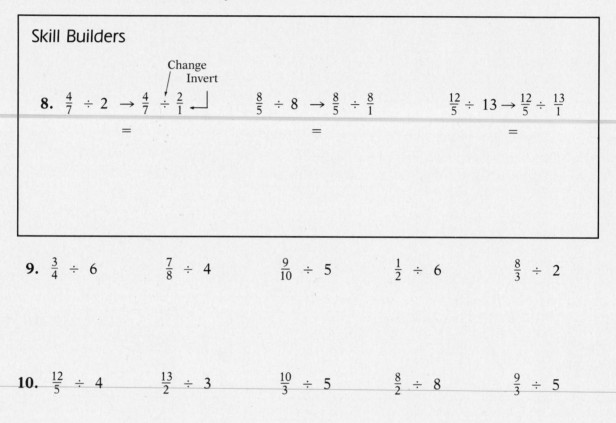

8. $\frac{4}{7} \div 2 \rightarrow \frac{4}{7} \div \frac{2}{1}$

$\frac{8}{5} \div 8 \rightarrow \frac{8}{5} \div \frac{8}{1}$

$\frac{12}{5} \div 13 \rightarrow \frac{12}{5} \div \frac{13}{1}$

=

=

=

9. $\frac{3}{4} \div 6$ $\frac{7}{8} \div 4$ $\frac{9}{10} \div 5$ $\frac{1}{2} \div 6$ $\frac{8}{3} \div 2$

10. $\frac{12}{5} \div 4$ $\frac{13}{2} \div 3$ $\frac{10}{3} \div 5$ $\frac{8}{2} \div 8$ $\frac{9}{3} \div 5$

You can also divide a whole number by a fraction. Again you write the whole number over the number 1, but this time you invert the fraction that is now the divisor.

Note: As the example shows, dividing a whole number or mixed number by a fraction can give a whole number answer.

Example:

$$6 \div \frac{2}{3} = \frac{6}{1} \div \frac{2}{3}$$

$$= \frac{6}{1} \times \frac{3}{2}$$

$$= \frac{18}{2} = 9$$

11. $5 \div \frac{1}{5} \rightarrow \frac{5}{1} \overset{\text{Change}}{\div} \frac{1}{5} \overset{\text{Invert}}{\rightarrow}$ $7 \div \frac{3}{4} \rightarrow \frac{7}{1} \div \frac{3}{4}$ $12 \div \frac{4}{3} \rightarrow \frac{12}{1} \div \frac{4}{3}$

$=$ $=$ $=$

12. $7 \div \frac{2}{3}$ $3 \div \frac{1}{4}$ $6 \div \frac{3}{8}$ $12 \div \frac{6}{2}$ $15 \div \frac{3}{2}$

Dividing with Mixed Numbers

When a division problem contains one or two mixed numbers, the first step is to change the mixed numbers to improper fractions. Then invert the divisor and multiply.

Change any improper fraction answer back to a mixed number.

Example 1:

$2\frac{1}{3} \div \frac{1}{2}$

$= \frac{7}{3} \div \frac{1}{2}$

$= \frac{7}{3} \times \frac{2}{1}$

$= \frac{14}{3}$

$= 4\frac{2}{3}$

Example 2:

$\frac{4}{5} \div 1\frac{2}{3}$

$= \frac{4}{5} \div \frac{5}{3}$

$= \frac{4}{5} \times \frac{3}{5}$

$= \frac{12}{25}$

13. $1\frac{3}{4} \div \frac{7}{8}$ $2\frac{1}{2} \div 1\frac{2}{3}$ $5 \div 3\frac{1}{2}$ $\frac{3}{5} \div 2\frac{3}{5}$

$= \frac{7}{4} \div \frac{7}{8}$ $= \frac{5}{2} \div \frac{5}{3}$ $= \frac{5}{1} \div \frac{7}{2}$ $= \frac{3}{5} \div \frac{13}{5}$

$= \frac{7}{4} \times \frac{8}{7}$ $= \frac{5}{2} \times \frac{3}{5}$ $=$ $=$

$=$ $=$ $=$ $=$

14. $2\frac{2}{3} \div \frac{2}{3}$ $3\frac{1}{2} \div 1\frac{1}{2}$ $6 \div 4\frac{1}{3}$ $4\frac{3}{5} \div 5$ $6\frac{3}{4} \div 2\frac{3}{8}$

15. $12\frac{1}{2} \div 6$ $5\frac{3}{4} \div 2$ $\frac{3}{8} \div 4$ $14 \div 3\frac{1}{3}$ $3\frac{1}{2} \div 1\frac{2}{3}$

Solving Multiplication and Division Word Problems

Fractions are used in a variety of both multiplication and division problems. As you work these problems, remember that your last step is to make sure that your answer makes sense.

Multiplication Word Problems

Most often, fraction multiplication problems are of two types:

- You are asked to find a **part** of a whole. In this type, the word *of* is often used to indicate multiplication.

    Example: What is $\frac{3}{4}$ *of* 12 miles?   **Answer:** $\frac{3}{4} \times \frac{12}{1} = 9$ miles

- You are given the size of one thing and asked to find the size of many.

    Example: If a small package weighs $\frac{1}{3}$ pound, how much do 7 identical packages weigh?   **Answer:** $\frac{1}{3} \times \frac{7}{1} = \frac{7}{3}$

    $= 2\frac{1}{3}$ **pounds**

Notice in each example that fraction multiplication is unlike whole number multiplication:

When you multiply a number by a fraction smaller than 1, your answer is a smaller number than the one you started with.

This is because you are finding a part of something. Use this fact to help check that each of your answers below makes sense.

Solve each multiplication problem below.

1. If "lean hamburger" is $\frac{1}{5}$ fat, how many ounces of fat are in a 12-ounce hamburger steak?

    (This is another way of asking, "What is $\frac{1}{5}$ of 12?")

2. Danny mixes $\frac{5}{6}$ pint of thinner in each gallon of stain he uses. How many pints of thinner will Danny need to complete a job that requires 18 gallons of stain?

3. Gloria uses $\frac{1}{4}$ of her monthly take-home pay to pay her rent. How much rent does she pay if her monthly take-home pay is $940?

    (This is another way of asking, "What is $\frac{1}{4}$ of $940?")

4. Frank exercises for $1\frac{1}{2}$ hours three days each week. At this rate, how many total hours of exercise does Frank get each week?

Division Word Problems

In a division word problem, you are most often trying to find how many of one object (the **part**) is contained in a second, larger object (the **whole**).

- The whole is the number that is being divided into. The whole must be written to the left of the division sign.

- The part is the number you are dividing by. The part must be written to the right of the division sign.

To divide, you invert the part and then multiply to find an answer. In each problem, be careful to correctly identify which number is the whole and which is the part.

Example 1: How many $\frac{1}{2}$-pound steaks can be made out of a package of hamburger that weighs $3\frac{1}{2}$ pounds?

Solution:

$$3\frac{1}{2} \div \frac{1}{2}$$

whole ↗   ↖ part

$$= \frac{7}{\cancel{2}_1} \times \frac{\cancel{2}^1}{1}$$

$$= \textbf{7 steaks}$$

Example 2: How many books that are $1\frac{1}{4}$ inches thick can be stacked between two shelves that are $13\frac{3}{4}$ inches apart?

Solution:

$$13\frac{3}{4} \div 1\frac{1}{4}$$

$$= \frac{55}{4} \div \frac{5}{4}$$

$$= \frac{\cancel{55}^{11}}{\cancel{4}_1} \times \frac{\cancel{4}^1}{\cancel{5}_1}$$

$$= \textbf{11 books}$$

Notice, as example 1 shows, fraction division is unlike whole number division:

When you divide a number by a fraction smaller than 1, your answer is a larger number than the one you started with.

Solve each division problem below.

5. The skateboard race covers a total of $1\frac{7}{8}$ miles. How many people will be needed on each team if each person skates for $\frac{3}{16}$ mile?

7. One centimeter is about $\frac{2}{5}$ inch. How many centimeters are in $\frac{7}{8}$ of an inch?

8. Bill has $\frac{7}{8}$ of a pizza that he wants to divide into 3 equal parts. What fraction of a whole pizza will be each person's share?

6. Losing weight at a rate of $3\frac{3}{4}$ pounds per month, how many months will it take Takeo to lose a total of $37\frac{1}{2}$ pounds?

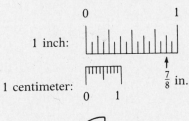

On these next two pages are groups of both multiplication and division problems. These problems are designed to increase your problem-solving skills with both fractions and mixed numbers.

Notice that the answers to problems 9, 10, and 11 are whole numbers. The way that these problems are worded alerts you that whole number answers are wanted.

9. Not counting any leftover piece, how many $12\frac{3}{4}$-inch pieces of ribbon can be cut from a piece that measures 141 inches in length?

10. For sale at the County Fair, Kelly prepared 1-pound bags of mixed candy. First she mixed sixteen $2\frac{2}{3}$-pound boxes of assorted candies together. Then she divided this mixture into 1-pound bags. If she measures carefully, how many full 1-pound bags can she make?

11. Bryan, a jewelry maker, uses $\frac{5}{16}$ ounces of gold for each custom-made ring he designs. How many complete rings can Bryan make when his gold supply is down to $3\frac{5}{8}$ ounces?

Problems 12 through 15 contain extra information. In each problem, circle the necessary information and then solve the problem.

12. An average American family spends about $\frac{1}{3}$ of its monthly net income on housing expenses, $\frac{1}{5}$ on food, $\frac{3}{20}$ on transportation costs, and $\frac{1}{10}$ on medical expenses. How much does a family with a monthly net income of $850 spend each month on food?

13. The distance around Jake's property is $3\frac{3}{4}$ miles. Across the back boundary it is $\frac{7}{8}$ mile. The local power company has decided to place power line poles along this back boundary. If the poles are to be placed $\frac{3}{64}$ mile apart, about how many will be placed along this boundary?

14. Last year, $\frac{3}{5}$ of the precipitation that fell in Klamath was rain, $\frac{3}{10}$ was snow, and $\frac{1}{10}$ was hail or sleet. If Klamath recorded 40 inches of precipitation during the year, how much of it fell as snow?

15. For the Saturday picnic, Laura bought $4\frac{1}{2}$ pounds of hamburger, $3\frac{3}{4}$ pounds of chicken, and 2 dozen hot dogs. If she plans to make $\frac{1}{4}$-pound hamburger patties, how many can she make with the meat she bought?

In problems 16 through 19, necessary information is found in the drawing to the right of each problem. Use this information to solve the problems.

16. Mandy divided a whole pizza as shown at right. If she and two friends plan to split the pizza equally, how many slices does each person get? Hint: Each gets a whole number of slices plus a fraction of another slice.

17. To bake cookies, Mark looked at a cookbook recipe. Part of that recipe is shown at the right. If Mark wanted to make only $\frac{1}{2}$ as many cookies as the recipe provides, how many cups of flour should he use?

Recipe
1 cup shortening
$1\frac{1}{2}$ cups sugar
$1\frac{3}{4}$ cups flour
1 cup raisins

18. Stella has to move the whole pile of sand shown at right. The maximum load she can carry each trip is $1\frac{2}{3}$ tons. Determine how many trips Stella will need to make in order to finish this job. Hint: the answer is a whole number!

sand   gravel

$16\frac{1}{4}$ tons $14\frac{1}{2}$ tons

19. On the engine Jim is fixing, the sparkplug "fires" when the top of the piston is $\frac{8}{9}$ of the way from the bottom to the top of the cylinder. Using the drawing at right, figure out how many inches above the bottom the piston top is when the plug fires.

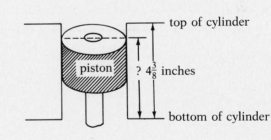

top of cylinder

piston   ? $4\frac{3}{8}$ inches

bottom of cylinder

Working with Decimals and Fractions at the Same Time

Decimals and fractions often appear in word problems together. Because of this, you will find it useful to be able to change common fractions to decimal fractions and, in some cases, to change decimal fractions to common fractions.

On the next few pages we'll discuss these skills and show you how they are used to solve some types of word problems.

Changing Common Fractions to Decimal Fractions

To change a common fraction to a decimal fraction, divide the denominator into the numerator. Then follow the rules for dividing decimal numbers.

Example: Change $\frac{3}{8}$ to a decimal fraction.

Step 1. Set up the problem for long division. Be sure to place the numerator 3 inside the division bracket.

Step 2. Place a decimal point to the right of the 3, add a zero, and divide.

Step 3. Continue dividing. You will need to add two more zeros before the answer comes out with no remainder.

$$
\begin{array}{r}
.375 \\
8\overline{)\,3.000} \\
\underline{2\,4} \\
60 \\
\underline{56} \\
40 \\
\underline{40} \\
0
\end{array}
$$

Answer: $\frac{3}{8}$ = **.375**

Change each fraction to a decimal. **Round** any **repeating decimal** to the hundredths place. (For a review of repeating decimals, see page 62, problem 16.) Complete the partially worked Skill Builders first.

Skill Builders

(repeating decimal)

1. $\frac{1}{4} \rightarrow 4\overline{)\,1.00}$ $\frac{5}{8} \rightarrow 8\overline{)\,5.000}$ $\frac{1}{6} \rightarrow 6\overline{)\,1.000}$

---

2. $\frac{2}{4}$ $\frac{3}{5}$ $\frac{3}{4}$ $\frac{7}{10}$ $\frac{3}{20}$

3. $\frac{1}{8}$ $\frac{1}{3}$ $\frac{6}{15}$ $\frac{2}{3}$ $\frac{5}{11}$

Changing Decimal Fractions to Common Fractions

To change a decimal fraction to a common fraction, write the number in the decimal as the numerator of a fraction. For the denominator of this fraction, write the place value of the final digit in the decimal. Reduce the fraction if you can.

Example 1: Write .075 as a fraction.

Step 1. Write 75 as the numerator of a fraction: $\frac{75}{}$

Step 2. Since 5 (of the .075) is in the thousandths place, one thousand will be the denominator of the fraction. $\frac{75}{1,000}$

Step 3. You can now reduce this fraction by dividing both top and bottom numbers by 25. $\frac{75 \div 25}{1,000 \div 25} = \frac{3}{40}$

Answer: $.075 = \frac{3}{40}$

Example 2: Write 2.45 as a mixed number.

Step 1. Write .45 as a fraction. $.45 = \frac{45}{100}$

Step 2. Reduce the fraction by dividing both top and bottom numbers by 5. $\frac{45 \div 5}{100 \div 5} = \frac{9}{20}$

Step 3. Replace .45 in the mixed decimal 2.45 by $\frac{9}{20}$ to make a mixed number.

Answer: $2.45 = 2\frac{9}{20}$

Write each of the following numbers as a fraction or mixed number. Reduce each fraction when possible.

4. .2   .5   .25   .75   .65

5. .95   .38   .125   .375   .750

6. 1.7   3.25   6.625   9.4   8.125

7. 6.005   7.035   3.0375   4.875   1.1375

Comparing Common Fractions with Decimal Fractions

To compare common fractions with decimal fractions, change the common fractions to decimals. This is by far the easiest way to make a comparison.

   With both numbers now expressed as decimals, follow the rules for comparing decimal fractions given on page 39.

Example: Which is larger, $\frac{3}{4}$ or .793?

Step 1.   Write $\frac{3}{4}$ as a decimal fraction. Dividing 4 into 3, we get $\frac{3}{4}$ = .75

$$\begin{array}{r} .75 = .750 \\ 4\overline{)3.00} \\ \underline{2\ 8} \\ 20 \\ \underline{20} \\ 0 \end{array}$$

Step 2.   Compare .75 with .793. Give each decimal the same number of places. Do this by adding 0 to the right of .75: .75 = .750.

   Since .793 is larger than .750, .793 is larger than .75 ($\frac{3}{4}$).

Answer: **.793 is larger than $\frac{3}{4}$.**

In each pair, circle the larger value.

1. .249 or $\frac{1}{4}$ $\frac{3}{5}$ or .62 $\frac{7}{8}$ or .859   .128 or $\frac{1}{8}$

To make each comparison below, round each fraction to 3 decimal places. Then circle the larger value in each pair.

2. .334 or $\frac{1}{3}$ $\frac{2}{7}$ or .283 $\frac{7}{9}$ or .804 $\frac{5}{6}$ or .812

3. Choose the smaller area:
   $\frac{3}{4}$ square yard or .74 square yard

4. Which represents the larger amount:
   $.71 or $\frac{5}{8}$ of a dollar?

5. Which is the heavier amount:
   4 ounces or .275 pound?

6. Which is the longer distance:
   25 inches or .7 yard?

Comparison Word Problems

A comparison word problem is solved by comparing one value with another. Most often, you are given information in both fraction and decimal form. You may be asked to find the difference in sizes, or you may be asked to write things in the order of their sizes.

To solve a comparison problem, rewrite each fraction as a decimal and then compare decimals as needed.

Solve each comparison problem below.

1. a) Will a wire that is $\frac{1}{8}$ inch wide fit through a hole that is .15 inch across?

   b) What is the difference in width between the wire and the hole?

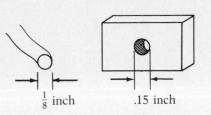

$\frac{1}{8}$ inch   .15 inch

2. According to the blueprint, Jim should drill a drainage hole "no larger than .24 inch in diameter (distance across)." If he has the three drill bit sizes to choose from as shown at right, which will give him the largest permissible hole?

Bit #	Diameter
#1	$\frac{1}{4}$ inch
#2	$\frac{7}{32}$ inch
#3	$\frac{3}{16}$ inch

3. Jill has four items to stack in a display case. She should place the heaviest item on the bottom, then the next heaviest, and so on. The lightest item goes on top. The weight of each item is listed at right.

   Place a letter on each blank line to show the order in which she should stack the items. The letter of the heaviest item goes on the bottom line.

Order	Item	Weight
_____	A	12 ounces
_____	B	1.2 pounds
_____	C	.78 pounds
_____	D	$\frac{4}{5}$ pound

4. Write the times shown at right as mixed decimals (for example, 6.30 hours). Express each answer to two decimal places, and write them in order. Write the shortest time at the left, and so on.

   _____ _____ _____ _____

Joyce: $6\frac{1}{3}$ hours

Irvin: 6.3 hours

Francis: 6 hours 22 minutes

Ellen: $6\frac{7}{20}$ hours

Solving Decimal *and* Fraction Word Problems

Many word problems (especially money problems) contain both decimals and fractions. To solve this type of problem, you may use either of two methods:

1. You can change all numbers to the same form, or

2. You can compute with the numbers as they are.

An example can best show each method.

Example: Mary Anne bought $5\frac{1}{4}$ yards of dress material for $2.48 per yard. What is the total cost of this material?

Method 1

Step 1. Change $5\frac{1}{4}$ to a decimal.

$5\frac{1}{4} = 5.25$ since $\frac{1}{4} = .25$

Step 2. Multiply $2.48 times 5.25.

```
    $2.48
  ×  5.25
   12 40
   49 6
  1240
 $1302 00  = $13.02
```

Method 2

Step 1. Change $5\frac{1}{4}$ to an improper fraction.

$5\frac{1}{4} = \frac{21}{4}$

Step 2. Multiply $2.48 times $\frac{21}{4}$.

$$\frac{\overset{62}{\cancel{\$2.48}}}{1} \times \frac{21}{\cancel{4}_{1}} = \$.62 \times 21 = \mathbf{\$13.02}$$

Using either method you prefer, solve the problems below.

1. While shopping at Myrta's Market, Jason bought $1\frac{3}{4}$ pounds of apples for $1.54. Figure out how much Jason paid for each pound of apples.

2. Last weekend, Hank worked $12\frac{1}{2}$ hours of overtime and was paid $8.70 per hour. How much did Hank earn for this overtime work?

3. At Sunday's race, Rhoda ran 3.5 miles in $27\frac{1}{2}$ minutes. On the average, how long did it take Rhoda to run each mile?

4. One-inch-wide wood trim costs $.28 per foot of length. How much does a piece of trim cost that is 6 feet 3 inches long? (Hint: First change 3 inches to a fraction of a foot.)

Solving Multi-Step Word Problems

Problems 1 through 5 are multi-step problems. In each a solution sentence is written and the missing information is underlined. Solve each problem by first finding the value of the missing information.

1. It takes Steve $\frac{1}{2}$ hour to drive to work each morning and $\frac{3}{4}$ hour to drive home each night. How much total time does Steve spend driving to and from work each 5-day week?

   total time = total driving time each day times 5

   $$\frac{1}{2} + \frac{3}{4}$$

   $$1\tfrac{1}{4} \times 5 = 6\tfrac{1}{4} \text{ hours}$$

2. Jay bought $5\frac{3}{8}$ pounds of chocolate. He gave $3\frac{1}{2}$ pounds of it to his sister. From the chocolate he kept, he made 12 chocolate desserts. How much chocolate did he put in each dessert?

   chocolate in each dessert = total chocolate kept divided by 12

3. A piece of tubing that is $71\frac{7}{8}$ inches long is to be cut into 3 smaller lengths, each $22\frac{1}{4}$ inches long. After the 3 pieces are cut off, what length of tubing will be left over?

   leftover length = $71\frac{7}{8}$ inches minus total length of the 3 smaller pieces

4. To prepare for a picnic, Kimberly bought two packages of hamburger: one marked 2 pounds 5 ounces and one marked 1 pound 7 ounces. Using both packages, how many $\frac{1}{4}$-pound hamburgers can she make? Hint: First add and then change ounces to a fraction of a pound: 16 ounces = 1 pound.

   number of hamburgers = total amount of meat divided by $\frac{1}{4}$

5. Monday through Friday, Rita earns her regular salary rate of $5.20 per hour. For each overtime hour she works on Saturday, she makes $1\frac{1}{2}$ times her regular rate. How much does Rita earn during a week in which she works $34\frac{1}{2}$ regular hours and $5\frac{3}{4}$ overtime hours?

   total earned = total regular pay plus total overtime pay

Solve problems 6, 7, and 8. You may find it helpful in each problem to first write a solution sentence to help identify missing information.

6. Rapid Express provides overnight mail delivery service. To mail packages, it charges $13.40 per pound. Using this service, what is the total cost of mailing three packages with weights of $1\frac{3}{8}$ pounds, $2\frac{1}{4}$ pounds, and $1\frac{1}{2}$ pounds?

7. For the hardware sale, Moses is going to make 15 equal-size bags of mixed screws, bolts, and nuts. First he'll mix $5\frac{3}{8}$ pounds of screws with $4\frac{1}{4}$ pounds of bolts and $3\frac{1}{2}$ pounds of nuts. Then he'll divide this mixture equally into 15 bags. About how much will each bag weigh?

8. June lives $1\frac{2}{3}$ miles from school. Two days each week she rides the bus to and from school. Three days each week she walks both ways. During each 5-day school week, how many miles does June walk going to and from school?

In problems 9 and 10, circle the arithmetic expression that will give the correct answer to each question. You do not need to solve these two problems.

9. For Halloween, Cindy bought a $9\frac{3}{8}$-pound bag of candy chews. She took out $\frac{3}{4}$ pound of the candy to keep for her own family. Then she divided the rest into 50 "treat bags" to give to children. About how much candy was in each treat bag?

a) $(9\frac{3}{8} \div 50) - \frac{3}{4}$

b) $(9\frac{3}{8} + \frac{3}{4}) \div 50$

c) $(50 - 9\frac{3}{8}) \times \frac{3}{4}$

d) $(9\frac{3}{8} - \frac{3}{4}) \div 50$

10. Martha bought $2\frac{1}{4}$ yards of plain material for $1.68 per yard, and $3\frac{1}{2}$ yards of print material for $2.46 per yard. How much did Martha pay for all this material?

a) $\frac{9}{4} \times 1.68 + \frac{7}{2} \times 2.46$

b) $(\frac{9}{4} + \frac{7}{2}) \times (1.68 + 2.46)$

c) $\frac{7}{2} \times 1.68 + \frac{9}{4} \times 2.46$

d) $(\frac{7}{2} + \frac{9}{4}) \times (2.46 - 1.68)$

Questions 11 through 13 refer to the following story.

Stock prices are given in dollars and fractions of dollars. The price quoted is the price of one share. For example, a stock that is quoted at $11\frac{1}{2}$ is selling for $11.50 per share.

In May, Sarah bought 15 shares of Best Computer stock, selling at $14\frac{3}{4}$. To buy the stock, Sarah paid the broker a commission of $11.35.

11. Including the commission, how much did Sarah pay for her 15 shares of Best Computer stock?

12. Between May and August, Best Computer stock rose to $17\frac{1}{2}$. How much did the price increase?

13. Sarah sold her stock when it was at $17\frac{1}{2}$. If she paid a commission of $13.60, how much money did she receive from the sale?

Questions 14, 15, and 16 refer to the story below.

To improve his health, Leon went on a special diet and exercise program. While on this program, Leon will lose $1\frac{3}{8}$ pounds each week. The exercise part of the program is as follows:

<u>1st 3 months:</u> Exercise $\frac{1}{2}$ hour per day, 3 days each week.

<u>2nd 3 months:</u> Exercise $\frac{3}{4}$ hour per day, 3 days each week.

<u>After 6 months:</u> Exercise $\frac{3}{4}$ hour per day, 4 days each week.

Leon's plan is to lose 66 pounds and decrease his weight from 256 pounds to 190 pounds.

14. While on this program, how much time will Leon spend exercising each week during the first three months?

15. How much more time per week will Leon spend exercising during the fourth month than he does during the first month?

16. How many weeks will it take Leon to reach 190 pounds?

Fraction Skills Review

On the next two pages, you'll have a chance to briefly review the main fraction computation skills. Check each answer with the answers given on page 193.

Comparing Values of Common Fractions:

Review page 71.

In each group of common fractions below, circle the two that are equal in value.

1. $\frac{1}{2}, \frac{5}{9}, \frac{4}{8}$ $\frac{3}{9}, \frac{2}{6}, \frac{3}{12}$ $\frac{5}{20}, \frac{3}{15}, \frac{6}{30}$ $\frac{3}{10}, \frac{6}{18}, \frac{9}{30}$

Arrange each group of fractions in order. Write the smallest to the left and the largest to the right. Use the largest denominator as the common denominator.

2. $\frac{5}{12}, \frac{1}{3}, \frac{1}{4}$ $\frac{13}{24}, \frac{5}{12}, \frac{5}{8}$ $\frac{1}{3}, \frac{7}{9}, \frac{2}{3}$ $\frac{9}{20}, \frac{5}{10}, \frac{2}{5}$

Adding and Subtracting Common Fractions:

Review pages 73 through 79.

3. $\frac{2}{4}$ $\frac{5}{8}$ $\frac{4}{7}$ $2\frac{3}{5}$ $6\frac{5}{8}$
   $+\frac{1}{4}$ $+\frac{1}{8}$ $+\frac{3}{7}$ $+\ \frac{1}{5}$ $+2\frac{7}{8}$

4. $\frac{3}{8}$ $\frac{7}{9}$ $1\frac{2}{3}$ $4\frac{3}{7}$ $7\frac{7}{8}$
   $+\frac{1}{4}$ $+\frac{2}{3}$ $+\ \frac{5}{6}$ $+2\frac{3}{14}$ $+5\frac{3}{4}$

5. $\frac{7}{9}$ $\frac{5}{8}$ $2\frac{3}{4}$ $6\frac{9}{16}$ $8\frac{3}{8}$
   $-\frac{2}{9}$ $-\frac{3}{8}$ $-\ \frac{1}{4}$ $-3\frac{5}{16}$ $-5\frac{1}{8}$

6. $\frac{7}{9} - \frac{2}{3}$ $\frac{10}{12} - \frac{1}{6}$ $5\frac{1}{4} - \frac{5}{8}$ $9\frac{7}{10} - 2\frac{2}{5}$ $13\frac{4}{9} - 7\frac{5}{6}$

Multiplying Common Fractions:

Review pages 92 through 95.

7. $\frac{2}{3} \times \frac{1}{5} =$ $\frac{3}{4} \times \frac{1}{2} =$ $\frac{3}{8} \times \frac{2}{5} =$ $\frac{7}{9} \times \frac{3}{4} =$

8. $\frac{3}{4} \times 7 =$ $\frac{2}{3} \times 9 =$ $5 \times \frac{6}{7} =$ $12 \times \frac{5}{6} =$

9. $\frac{1}{2} \times 1\frac{1}{3} =$ $2\frac{1}{4} \times \frac{2}{3} =$ $5 \times 3\frac{5}{8} =$ $3\frac{4}{5} \times 5 =$

10. $3\frac{3}{5} \times 4\frac{2}{3} =$ $5\frac{3}{8} \times 2\frac{1}{4} =$ $\frac{7}{3} \times 4\frac{3}{7} =$ $6\frac{3}{8} \times \frac{9}{2} =$

11. $\frac{1}{2} \times \frac{3}{5} \times \frac{10}{15} =$ $\frac{2}{3} \times \frac{3}{4} \times 5 =$ $2\frac{1}{2} \times \frac{3}{4} \times \frac{4}{5} =$

Dividing Common Fractions: Review pages 96 through 99.

12. $\frac{3}{4} \div \frac{1}{2} =$ $\frac{5}{8} \div \frac{1}{3} =$ $\frac{1}{6} \div 3 =$ $4 \div \frac{3}{8} =$

13. $5 \div \frac{3}{10} =$ $\frac{2}{3} \div 6 =$ $\frac{3}{4} \div 3\frac{1}{3} =$ $4\frac{3}{8} \div \frac{1}{4} =$

14. $5\frac{3}{5} \div \frac{2}{3} =$ $6 \div 1\frac{2}{3} =$ $4\frac{3}{7} \div 5 =$ $1\frac{7}{8} \div 4 =$

15. $2\frac{2}{5} \div 1\frac{2}{3} =$ $4\frac{1}{2} \div 1\frac{3}{7} =$ $6\frac{1}{3} \div 7\frac{4}{9} =$ $5\frac{5}{6} \div 9\frac{1}{2} =$

5
Percent Skills

Percent is another way to write part of a whole. But, because percent means hundredth, **a percent always refers to a whole that is divided into 100 equal parts.**

You write percent as the number of hundredths followed by the percent sign %. For example, 10 percent is written as 10%. **10% means 10 parts out of 100.**

Any percent can easily be written as an equivalent decimal or fraction.

35 ☐

100 equal parts

- Percent has the same value as a two-place decimal.
  For example, 35% is equal to .35.

- Percent has the same value as a fraction that has a denominator of 100. For example, 35% is equal to $\frac{35}{100}$.

35% is shaded.

Since percents, decimals, and fractions are so similar, you may wonder why we use all three! Wouldn't one do just fine? The answer is yes. But, like so many other things, math has a history. As math developed, each method of writing part of a whole became popular for certain uses:

- Decimals became the basis of our money system and of the metric measuring system.

- Fractions became the basis of the English measuring system.

- Percents became used with such things as store sales, taxes, finance charges, and rates of increase and decrease.

In the pages ahead, you'll see that percents, decimals, and fractions often appear together. Don't let this confuse you. **You'll always change a percent to a decimal or fraction before you do any computations.**

To remember the meaning of percent, many students just look at the % sign.

- When you want to write a percent as a decimal, the two zeros in % remind you of two decimal places.   35% = .35

- When you want to write a percent as a fraction, the two zeros in % remind you of two zeros in the denominator.   27% = $\frac{27}{100}$

In each figure below, show how the shaded part can be written as a **fraction**, as a **decimal**, and as a **percent**. The first problem is done as an example.

1. $\frac{21}{100}$
.21
21%

2. _____ _____

3. _____ _____

4. _____ _____

5. _____ _____

6. _____ _____

7. 100% always represents a whole amount. At right, 30% of the large square is shaded; 70% is unshaded. Notice that 30% + 70% = 100%, the whole square.

% shaded   30%
% unshaded   70%
total %   100%

Write what percent of each square below is shaded, what percent is unshaded, and what the total of the two percents is equal to.

a) % shaded   _____
        % unshaded _____
        total %   _____

b) % shaded   _____
        % unshaded _____
        total %   _____

8. One hundred students attend GED classes at Sheldon Community Center. Eighty-nine students showed up for classes on Thursday.

  a) What percent of students came to class Thursday?

  b) What percent of students did not come to class?

9. A meter is a metric unit of length that is a little longer than a yard. A meter is divided into 100 smaller units called *centimeters*. What percent of 1 meter is a length of 57 centimeters?

10. There are 100 pennies in 1 dollar. What percent of 1 dollar is each of the following amounts?

  a) penny?   c) a dime?
  b) a nickel? d) a quarter?

11. Vern is entered in a 100- mile bike race. How many miles must Vern ride to complete 75% of the race?

115

Percents, Decimals, and Fractions

To become comfortable with percents, decimals, and fractions, you'll want to learn how you can quickly change from one form to another. This skill is very important. You'll use it often in word problems, especially in those where percents, decimals, and fractions appear together in the same problem.

On the next few pages, work carefully to develop this skill. We'll start by seeing how to change percents to decimals.

Changing Percents to Decimals

To change a percent to a decimal,

- move the decimal point two places to the left, adding a zero if necessary;
- drop the percent sign;
- drop any unnecessary zeros.

Remember that the decimal point in a whole number percent is understood to be at the right of the whole number even though it is not written.

Examples:

Percent	Move Decimal Point Two Places Left	Decimal	
35%	₂35	.35	
80%	₂80	.80 = .8	←The zero can be dropped.
9%	₂09 ←Add a zero.	.09	
8.5%	₂08.5	.085	
62.5%	₂62.5	.625	

Change each percent below to a decimal. Add zeros where necessary to give *at least* two decimal places. Drop any unnecessary zeros.

1. 55% =   25% =   10% =   27% =

2. 7% =   4% =   1% =   8% =

3. 5.5% =   6.6% =   8.8% =   11.5% =

4. 30% =   7.5% =   75% =   3% =

Changing Fractional Percents to Decimals

A fractional percent is a percent less than 1%. A fractional percent can be written alone such as $\frac{1}{2}\%$ or it may be part of a mixed number percent such as $33\frac{1}{3}\%$.

To change a fractional percent to a decimal, first use division to change the fraction itself to a decimal. This changes the fractional percent to a decimal percent. Next, move the decimal point two places to the left to change the decimal percent to a decimal fraction.

Examples:

Fractional Percent		Decimal Percent		Decimal Fraction	Graphically
a) $\frac{1}{2}\%$	=	.5%	=	.005	

Caution: Don't make the mistake of writing $\frac{1}{2}\% = .5$, a commonly made error!

$\frac{1}{2}\%$ is shaded

$\frac{1}{2}\%$ is $\frac{1}{2}$ of 1%

| b) $\frac{1}{4}\%$ | = | .25% | = | .0025 |

$\frac{1}{4}\%$ is shaded

$\frac{1}{4}\%$ is $\frac{1}{4}$ of 1%

| c) $47\frac{4}{5}\%$ | = | 47.8% | = | .478 |

$47\frac{4}{5}\%$ is shaded.

Note: The graphics at right help you visualize just how small fractional percents really are.

Complete the row of partially worked Skill Builders. Then change each percent below to a decimal fraction.

Skill Builders

5. $\frac{1}{5}\% = .2\% =$ $\qquad$ $\frac{7}{10}\% = .7\% =$ $\qquad$ $3\frac{1}{2}\% = 3.5\% =$

6. $\frac{3}{10}\% =$ $\qquad$ $\frac{2}{5}\% =$ $\qquad$ $\frac{3}{4}\% =$ $\qquad$ $\frac{1}{2}\% =$

7. $7\frac{3}{4}\% =$ $\qquad$ $8\frac{1}{2}\% =$ $\qquad$ $9\frac{3}{8}\% =$ $\qquad$ $4\frac{3}{5}\% =$

Changing Percents Larger Than 100% to Decimals

Once in a while you may see percents that are larger than 100% (larger than a whole unit). When you change such a large percent to a decimal, the answer will be a whole number or a mixed decimal.

Examples: 1. 100% = 1.00   = **1**   2. 300% = 3.00 = **3**
            3. 250% = 2.50 = **2.5**   4. 575% = 5.75 = **5.75**

Change each percent below to a whole number or a mixed decimal.

8. 400% =   200% =   700% =   900% =

9. 350% =   485% =   130% =   225% =

In each large square below, shade the number of smaller squares that represents each indicated percent. Each large square (1 whole unit) contains 100 equal-size smaller squares.

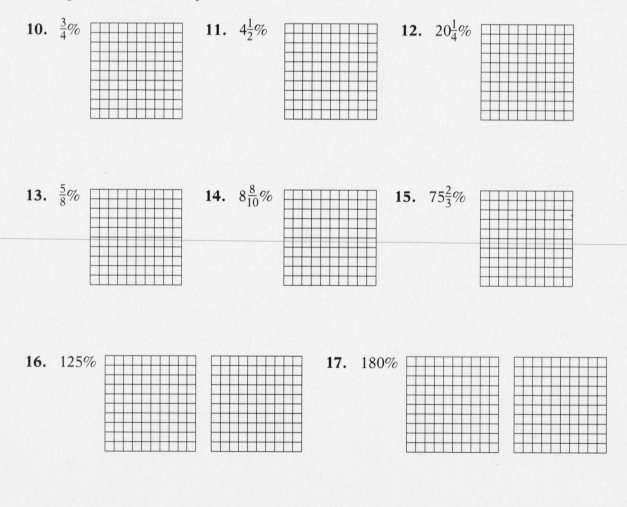

10. $\frac{3}{4}$%   **11.** $4\frac{1}{2}$%   **12.** $20\frac{1}{4}$%

13. $\frac{5}{8}$%   **14.** $8\frac{8}{10}$%   **15.** $75\frac{2}{3}$%

16. 125%   **17.** 180%

118

Changing Decimals to Percents

To change a decimal to a percent, simply reverse the rules for changing a percent to a decimal:

- move the decimal point two places to the right, adding a zero if necessary;
- write the percent sign after the number;
- drop any unnecessary zero;
- drop the decimal point unless the percent contains a decimal fraction.

Examples:

Decimal	Move Decimal Point Two Places Right	Percent
.25	.25.	25% ← Drop the point.
.90	.90.	90%
.07	.07.	7% ← Drop the unnecessary zero.
.4	.40. ← Add a zero.	40%
.125	.12.5	12.5%
2.5	2.50. ← Add a zero.	250%

Change each decimal below to a percent.

18. .75 =   .50 =   .65 =   .8 =

19. .09 =   .02 =   .375 =   .085 =

20. 1.5 =   .605 =   2.25 =   .7 =

Divide as indicated. Change each decimal answer to a percent.

21. $9 \div 10 =$ $1 \div 2 =$ $3 \div 4 =$ $7 \div 8 =$

22. $4 \div 5 =$ $3 \div 8 =$ $5 \div 10 =$ $8 \div 20 =$

23. $8 \div 5 =$ $14 \div 4 =$ $24 \div 6 =$ $9 \div 3 =$

Changing Percents to Fractions

To change a percent to a fraction: 1) write the percent as a fraction with 100 as the denominator, 2) reduce the fraction if possible.

Example 1: Change 45% to a fraction.

Step 1. Write the percent (45) as a fraction with 100 as the bottom number.

$$45\% = \frac{45}{100}$$

Step 2. Reduce the fraction by dividing top and bottom by 5.

$$\frac{45 \div 5}{100 \div 5} = \frac{9}{20}$$

Answer: $\frac{9}{20}$

Notice that we can also change a percent that's larger than 100% to a mixed number.

Example 2: Change 225% to a mixed number.

Step 1. Write the percent (225) as a fraction with 100 as the denominator.

$$225\% = \frac{225}{100}$$

$$2\frac{25}{100} = 2\frac{1}{4}$$

Step 2. Divide 225 by 100. Then reduce the proper fraction remainder.

$$100)\overline{225}$$
$$\underline{200}$$
$$25$$

Answer: $2\frac{1}{4}$

Change each percent below to a fraction.

24. 55% =   40% =   15% =   8% =

25. 1% =   30% =   5% =   50% =

26. 12% =   99% =   20% =   6% =

27. 250% =   475% =   125% =   340% =

Changing Fractions to Percents

To change a fraction to a percent: 1) reduce the fraction, 2) change the fraction to a decimal, 3) change the decimal to a percent.

Example: 1 Change $\frac{3}{4}$ to a percent.

Change $\frac{3}{4}$ to a decimal. Then change the decimal to a percent.

$$.75 = \mathbf{75\%}$$
$$4\overline{)3.00}$$
$$\underline{2\ 8}$$
$$20$$
$$\underline{20}$$

Example 2: Change $\frac{3}{8}$ to a percent.

Change $\frac{3}{8}$ to a decimal and then to a percent.

$$.375 = \mathbf{37.5\%}$$
$$8\overline{)3.000}$$
$$\underline{2\ 4}$$
$$60$$
$$\underline{56}$$
$$40$$
$$\underline{40}$$

A **shortcut** that's easy to remember is just to multiply the fraction by 100%. Look how this shortcut simplifies the examples above:

$$\overset{1}{\underset{1}{\cancel{\frac{3}{4}}}} \times \overset{25\%}{\cancel{\frac{100\%}{1}}} = \frac{75\%}{1} = 75\% \qquad \frac{3}{8} \times \frac{100\%}{1} = \frac{300\%}{8} = 37.5\%$$

Using either method, change each fraction below to a percent.

28. $\frac{2}{4} =$ $\frac{3}{5} =$ $\frac{1}{4} =$ $\frac{3}{6} =$

29. $\frac{1}{8} =$ $\frac{3}{10} =$ $\frac{1}{3} =$ $\frac{1}{6} =$

30. $\frac{7}{25} =$ $\frac{5}{9} =$ $\frac{9}{10} =$ $\frac{7}{8} =$

31. $\frac{11}{20} =$ $\frac{1}{5} =$ $\frac{2}{3} =$ $\frac{3}{16} =$

Commonly Used Percents, Decimals, and Fractions

You have now seen how percents, decimals, and fractions can be changed from one to another. This skill will be used throughout the word problems on the pages ahead.

Below is a partially completed chart of the 33 most commonly used percents, decimals, and fractions. The decimals and fractions are written in their most reduced form. Complete this chart by filling in the blank spaces. It is a good idea to try to memorize the completed chart. Doing so will save you time in future work. Be sure to check your answers before you go on.

Percent	Decimal	Fraction
10%	.1	$\frac{1}{10}$
	.2	
		$\frac{1}{4}$
30%		
40%		
	.5	

Percent	Decimal	Fraction
	.6	$\frac{3}{5}$
70%		
	.75	
		$\frac{4}{5}$
	.9	

There are two special fractions that we have not yet mentioned. These are $\frac{1}{3}$ and $\frac{2}{3}$. They are special because their decimal and percent forms include a fraction. This is because $\frac{1}{3}$ and $\frac{2}{3}$ are **repeating decimals**. This simply means that when you write either $\frac{1}{3}$ or $\frac{2}{3}$ as a decimal or percent, you always end up with a remainder.

Because the fractions $\frac{1}{3}$ and $\frac{2}{3}$ often appear in word problems, **you should memorize the three forms of each given below**.

Percent	Decimal	Fraction
$33\frac{1}{3}\%$	$.33\frac{1}{3}$	$\frac{1}{3}$
$66\frac{2}{3}\%$	$.66\frac{2}{3}$	$\frac{2}{3}$

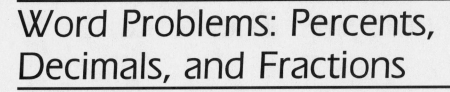

Word Problems: Percents, Decimals, and Fractions

Answer each question below. Remember to reduce fractions to lowest terms.

1. By 9:00, Mary had completed 60% of her homework.

   a) By 9:00, what fraction of her homework had Mary done?

   b) How is 60% written as a decimal?

2. Bert went to a sale where "all clothes are 25% off."

   a) Express this price reduction as a decimal.

   b) Express this price reduction as a fraction.

3. As of June 15, Mathew had completed $66\frac{2}{3}$% of the house he was building.

   a) What fraction of the house had Mathew completed?

   b) What percent of the house remains to be built?

4. A dollar contains 20 nickels.

   a) What fraction of a dollar is 13 nickels?

   b) What percent of a dollar is 13 nickels?

5. Joan rode in a 100-mile bike race.

   a) What percent of the race had she finished when she had gone 62.5 miles?

   b) At the 62.5-mile point, what percent of the race did she still have left?

6. A kilometer is a metric distance unit that equals 1,000 meters.

   a) Expressed as a decimal, what part of a kilometer is 350 meters? (Hint: 350 meters is 350 thousandths of a kilometer.)

   b) What percent of a kilometer is 350 meters?

7. A ton is equal to 2,000 pounds.

   a) What fraction of a ton is 1,500 pounds?

   b) What percent of a ton is 1,500 pounds?

Using Percents: The Percent Circle

In the rest of this chapter, we're going to look at the three ways percent is used in calculations:

- To find **part** of a whole
  Example: What is 10% of 200 pounds?

- To find what **percent** a part is of a whole
  Example: What percent of 300 feet is 60 feet?

- To find a **whole** when a part of it is given
  Example: If 20% of a bill is $45, how much is the whole bill?

The Percent Circle

Learning to be able to find a **part,** a **percent**, or a **whole** in a wide variety of problems is the goal of the rest of this chapter. To help you learn and remember how to solve each type of problem, we'll use a memory aid called the **percent circle**. Memorize this circle and its uses on the pages ahead.

PERCENT CIRCLE

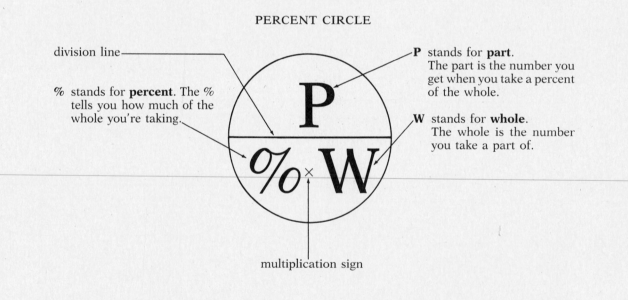

division line

% stands for **percent**. The % tells you how much of the whole you're taking.

P stands for **part**. The part is the number you get when you take a percent of the whole.

W stands for **whole**. The whole is the number you take a part of.

multiplication sign

Example 1: Identify the %, W, and P in sentence *a* below:

a) 15% of 300 is 45

Answer:
% = 15%
W = 300
P = 45

Example 2: Identify the %, W, and P in sentence *b* below:

b) 50 is 25% of 200

Answer:
% = 25%
W = 200
P = 50

Using the Percent Circle

You can use the percent circle to remember both multiplication and division problems that involve percents. To use the percent circle, you simply cover the number you're trying to find. This unknown number is then computed by doing the calculation that the uncovered symbols tell you to do. Since there are three different symbols, we can write three different **percent sentences** as shown below. We'll use each of these percent sentences in the pages ahead.

I. To find **part** of a whole:

> *Step 1.* Place your finger over the P (part), the number you're tying to find.
>
> *Step 2.* Read the uncovered symbols: $\% \times W$.
> Write: $P = \% \times W$

Percent Sentence I: **To find the part, you multiply the percent by the whole.**

II. To find what **percent** a part is of a whole:

> *Step 1.* Place your finger over the % (percent), the number you're trying to find.
>
> *Step 2.* Read the uncovered symbols: $\frac{P}{W}$
> Write: $\% = \frac{P}{W}$

Percent Sentence II: **To find the percent, you divide the part by the whole.**

Notice that the line that crosses the center of the percent circle is a division line.

III. To find a **whole** when a part of it is given.

> *Step 1.* Place your finger over the W (whole), the number you're trying to find.
>
> *Step 2.* Read the uncovered symbols: $\frac{P}{\%}$
> Write: $W = \frac{P}{\%}$

Percent Sentence III: **To find the whole, you divide the part by the percent.**

Complete the percent circle and the three sentences partially completed below.

1. To find the **part**, _____.
2. To find the **percent**, _____.
3. To find the **whole**, _____.

Finding Part of a Whole

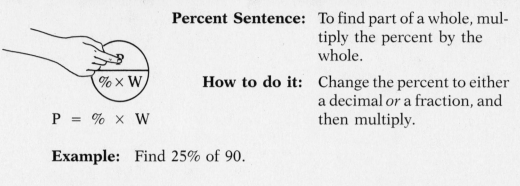

$P = \% \times W$

Percent Sentence: To find part of a whole, multiply the percent by the whole.

How to do it: Change the percent to either a decimal *or* a fraction, and then multiply.

Example: Find 25% of 90.

Method 1

Step 1. Change 25% to a decimal.

$$25\% = .25$$

Step 2. Multiply 90 by .25.

$$\begin{array}{r} 90 \\ \times .25 \\ \hline 4\,50 \\ 18\,0 \\ \hline 22.50 = 22.5 \end{array}$$

Answer: 22.5 (or $22\frac{1}{2}$)

Method 2

Step 1. Change 25% to a fraction.

$$25\% = \frac{25}{100} = \frac{1}{4}$$

Step 2. Multiply 90 by $\frac{1}{4}$.

$$\frac{1}{4} \times \frac{90}{1} = \frac{90}{4}$$
$$= 22\frac{2}{4} = 22\frac{1}{2}$$

Answer: $22\frac{1}{2}$ (or 22.5)

Since both methods give the same answer, you can use either method. But, when the percent is $33\frac{1}{3}\%$ (= $\frac{1}{3}$) or $66\frac{2}{3}\%$ (= $\frac{2}{3}$), you should always use method 2.

As a review, change each percent in row 1 to either a decimal or a fraction.

1. 50%   34%   125%   75% $6\frac{1}{2}\%$

Solve each problem below. The first problem in each row is worked as an example.

Percents Between 1% and 100%

2. 30% of 50   10% of 80   1% of 220 $33\frac{1}{3}\%$ of 189

$$30\% = .30 \quad \begin{array}{r} 50 \\ \times .30 \\ \hline 15.00 \end{array}$$

3. 25% of 17   45% of 30   50% of 121   7% of 42

$$25\% = \frac{1}{4}$$
$$\frac{1}{4} \times \frac{17}{1} = \frac{17}{4} = 4\frac{1}{4}$$
$$(\text{or } 4.25)$$

126

Decimal and Fractional Percents

4. .5% of 12   .4% of 14   .25% of 72   .75% of 10

$.5\% = .005$

$$\begin{array}{r} 12 \\ \times .005 \\ \hline .060 \end{array}$$

5. 8.8% of 130   5.5% of 225   7.5% of 175   11.5% of 50

$8.8\% = .088$

$$\begin{array}{r} 130 \\ \times .088 \\ \hline 1040 \\ 1040 \\ \hline 11.440 \end{array} = 11.44$$

6. $\frac{3}{4}$% of 40 $\frac{1}{2}$% of 60 $\frac{1}{4}$% of 100 $\frac{3}{5}$% of 15

$\frac{3}{4}\% = .75\% = .0075$

$$\begin{array}{r} 40 \\ \times .0075 \\ \hline 200 \\ 280 \\ \hline .3000 \end{array} = .3$$

7. $2\frac{1}{2}$% of 50 $8\frac{1}{2}$% of 750 $5\frac{1}{4}$% of 92 $4\frac{3}{4}$% of 30

$2\frac{1}{2}\% = 2.5\% = .025$

$$\begin{array}{r} 50 \\ \times .025 \\ \hline 250 \\ 100 \\ \hline 1.250 \end{array} = 1.25$$

Percents Larger than 100%

8. 200% of 40   300% of 64   500% of 130   400% of 19

$200\% = 2.00 = 2$

$$\begin{array}{r} 40 \\ \times\ 2 \\ \hline 80 \end{array}$$

9. 125% of 26   240% of 64   150% of 48 $166\frac{2}{3}$% of 12

$125\% = 1.25$ $(166\frac{2}{3}\% = 1\frac{2}{3})$

$$\begin{array}{r} 26 \\ \times 1.25 \\ \hline 130 \\ 52 \\ 26 \\ \hline 32.50 \end{array} = 32.5$$

Word Problems: Finding Part of a Whole

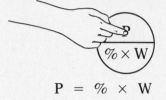

$$P = \% \times W$$

Percent word problems often ask you to find part of a whole amount. This also is a one-step problem in which you change the percent to a decimal or to a fraction. Then you multiply to determine the answer.

Example: Lowrey's Real Estate charges a 6% sales commission for each house it sells. When Jan Irving had Lowrey's sell her house, a sales price of $58,400 was agreed upon. How much commission will Lowrey's charge for the sale of Jan's house?

To determine the commission, change 6% to .06 and multiply.

Identify % and W
% = 6%
W = $58,400
P = unknown commission

$58,400	selling price (W)
× .06	× commission rate (%)
$3504.00	commission (P)

Answer: **$3,504**

Solve each problem below. First identify the percent and the whole.

1. Out of his monthly salary of $1,246, Greig puts 15% in a savings account. At this rate, how much does Greig save each month?

2. During last year, the inflation rate on consumer goods averaged only 4%. How much did the cost of a $285 washing machine increase during this year?

3. To pass her math test, Belinda must correctly answer 65% of the questions. What number of questions must she correctly answer in order to pass a test that contains 80 questions?

4. Only 23% of the registered voters in Benton County voted in last week's budget election. How many people voted if there are 74,000 registered voters in this county?

5. Because of a company cutback, Randi's salary of $385 a week was cut by 9%. Figure out how much less money Randi will earn each week after the cutback.

Word Problems: Increasing or Decreasing a Whole

$P = \% \times W$

Many word problems involve increasing a whole by **adding** part of its value. Other problems involve decreasing a whole by **subtracting** part of its value. To work these types of problems, you first find the unknown part (the amount of increase or decrease). Then you add or subtract this part from the original whole.

A common example of increasing a whole by adding a part is the sales tax. Other examples are given in the problems below.

SALES TAX: Purchase price   =   selling price   +   amount of sales tax

Example: A pair of jogging shoes is marked with a selling price of $46. What is the purchase price if there is a 6% state sales tax?

$\% = 6\%$
$W = \$46$
$P = $ unknown amount of tax
Purchase price $= W + P$

Step 1.   Find the amount of the sales tax. Change 6% to .06 and multiply.

$$\begin{array}{r} \$46 \\ \times\ .06 \\ \hline \$2.76 \end{array}$$

selling price (W)
$\times$ sales tax rate (%)
amount of tax (P)

Step 2.   Add the sales tax to the price of the shoes. This will give the purchase price.

$$\begin{array}{r} \$46.00 \\ +\quad 2.76 \\ \hline \$48.76 \end{array}$$

selling price (W)
$+$ amount of tax (P)
purchase price

Answer: $48.76

The following problems involve increasing a whole by adding a part of its value.

RATE INCREASE: New amount = original amount + amount of increase

1. Before his raise, Jose made a monthly salary of $860. What is the amount of Jose's new salary if he was given a 5% raise?

2. Last year during April, the Ortega family used 1,500 kilowatt-hours of electricity. Now, with a new baby, they used 16% more this April. How many kilowatt-hours of electricity did the Ortegas use this April?

MARKUP: Selling price = store's cost + markup

3. In Cecil's Mens Store, Cecil pays $12.00 for the Fashion Plus shirts that he sells. If he adds a 30% markup to his cost, for what price does Cecil sell Fashion Plus shirts?

4. Discount Auto Parts adds a markup of only 18% on every item that they sell. To make this much profit, for what price should Discount Auto sell tire chains that cost the store $16.50?

The following problems involve decreasing a whole by subtracting part of its value.

DISCOUNT: Sale price = original price − amount of discount

5. Rosa's Dress Shop is having a sale. All dresses are to be sold "30% off." At this sale, what will be the price of a dress that originally sold for $32.00?

6. As a special deal, Beth's friend offered to sell her a new TV at a discount of 40%. How much would Beth have to pay for the set if its regular price was $640?

RATE DECREASE: New amount = original amount − amount of decrease

7. In order to help the company, employees of Thomas Electronics agreed to a 10% pay cut. What will Shirley's new salary be if before the cut she was earning $746 per month?

8. Due to reducing the city's speed limit to 30 mph, Redville had a 15% decrease in traffic accidents during May this year compared to last year. How many May accidents did Redville have this year if last May it had 40?

DEPRECIATION: New value = original value − amount of depreciation

Note: Depreciation is value lost due to age or wear.

9. A new car depreciates (loses value) 20% during its first year. Much of this loss in value takes place the moment the car is driven off the new car lot because it is no longer a new car! What will be the approximate value of Ellie's new $8,350 car one year after her purchase?

10. During a time of high unemployment in his hometown, Jess saw the value of his house depreciate 12% in one year. What was his house worth at the end of the year if its value at the beginning of the year was $64,000?

130

Word Problems: Mixed Practice

Solve each problem below.

1. Benny lives in a state that has a 5% sales tax. What price will he have to pay for a suit that is marked at a sale price of $149.50?

2. Bill's Discount Realty charges a real estate commission of only 4%. If Bill sells a house for a sales price of $72,000, how much commission will he earn?

3. Cindy owns and operates "Cindy's Cuts," a hair salon and beauty shop. On each hair product she sells, Cindy adds a markup of 25%. What price does Cindy charge for shampoo that costs her $4.80?

4. Certain machines depreciate at a rate of about 15% each year. At the beginning of his third year of business, Daniel figured that his tractor was worth about $38,000. What value would this tractor have at the end of that third year?

5. While shopping for clothes, Sally saw a shaker sweater marked as "40% off." Determine the amount of the sale price if the original price was listed as $42.50.

In problems 6 and 7, circle the arithmetic expression that will give the correct answer to each question. You do not need to solve these two problems.

6. During the last 3 months, the number of students in Lee's GED class has increased by 10%. What number of students are in the class now if 3 months ago there were 20?

a) $(.10 \times 20) + 20$
b) $20 \div .10 + 20$
c) $(.10 \times 20) - 20$
d) $20 - .10 \times 20$

7. The population of Orin has decreased by $33\frac{1}{3}\%$ over the past ten years. Ten years ago 38,700 people lived there. What's the town population now?

a) $38,700 + (\frac{1}{3} \times 38,700)$
b) $(\frac{1}{3} \times 38,700) + 10,000$
c) $38,700 - (\frac{1}{3} \times 38,700)$
d) $\frac{1}{3} \times 38,700$

Finding What Percent a Part Is of a Whole

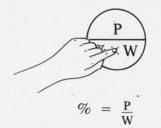

$$\% = \frac{P}{W}$$

Percent Sentence: To find what percent a part is of a whole, divide the part by the whole.

How to do it: Write a fraction $\frac{P}{W}$. Reduce this fraction, divide, and change the quotient to a percent.

Example: 8 is what percent of 32?

Step 1. Write the fraction 8 (P) over 32 (W). Reduce this fraction.

$$\frac{P}{W} = \frac{8}{32} = \frac{1}{4}$$

Step 2. Change $\frac{1}{4}$ to a decimal by dividing 4 into 1:
$$\frac{1}{4} = .25$$

Step 3. Change the decimal .25 to a percent by moving the decimal point two places to the right and adding a % sign.

$$.25 = 25\%$$

Answer: 8 is **25%** of 32

Note: Another way to ask the same question is to write, "What percent of 32 is 8?" In each case, 32 (the number following the word *of*) is considered to be the whole (W) and is written as the denominator of the fraction.

As you solve each problem below, remember to first write $\frac{P}{W}$ as a fraction. Then reduce this fraction if possible before dividing.

1. 5 is what percent of 25?

2. 10 is what percent of 40?

3. 8 is what percent of 16?

4. What percent of 120 is 24?

5. What percent of 60 is 15?

6. What percent of 16 is 2?

The answers to problems 7 and 8 are larger than 100%.

7. 250 is what percent of 125?

8. What percent of 30 is 90?

Word Problems: Finding What Percent a Part Is of a Whole

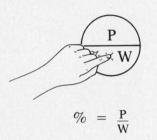

$$\% = \frac{P}{W}$$

Before writing the fraction $\frac{P}{W}$ you need to identify which number is the part (P) and which is the whole (W).

In almost all word problems you'll work, P is the smaller number and goes on top. However, just to be careful, rephrase the question in each problem as follows: "____ is what percent of ____?" The number that you choose to follow the word *of* will be W.

Solve each problem below. As a first step in problems 1, 2, and 3, fill in the blanks of the rephrased question below each problem. Remember, always reduce a fraction before dividing. Reducing makes your work much easier.

1. The Zarate family makes $240 each week. If they spend $60 each week on food, what percent of their income goes for food?

   __60__ is what percent of __240__?

2. What percent of a yard is 27 inches? (1 yard = 36 inches)

   _____ is what percent of _____?

3. During last year, the price of a $250 washing machine increased by $25. What percent of an increase in price is this?

   _____ is what percent of _____?

4. On her science test, Jamie got 40 questions correct out of a total of 60 questions. What percent of correct answers did Jamie get? (Hint: The answer will contain a fraction.)

5. Meyer's Department Store pays $8.00 for each shirt it sells. If each shirt is priced so that the store makes a $4.00 profit, what percent markup does Meyer's use?

6. Two years after she bought a used Chevrolet for $5,500, its value had decreased by $1,760. What percent of depreciation is this over the two-year period?

7. Out of Gina's gross monthly pay of $900, her employer withholds $171 for federal tax. What percent of Gina's salary is withheld for this tax?

Word Problems: Finding Percent Increase or Decrease

One type of two-step word problem that you'll often see involves finding a percent increase or a percent decrease. In these problems, you divide the **change in an amount** by the **original amount** (the whole). The percent circle is used in the second step of the solution.

Example: At a clothing sale, the price of a wool skirt was reduced from $48 to $36. What percent price reduction is this?

Step 1. Subtract to find how much the skirt has been reduced in price:
$48 − $36 = $12.
$12 is the **amount of decrease**.

Step 2. Ask yourself, "What percent of $48 is $12?"
From the percent circle, write
$$\% = \frac{P}{W} = \frac{\$12}{\$48} = \frac{1}{4}$$

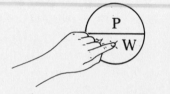

$$\% = \frac{P}{W} = \frac{\text{amount of decrease}}{\text{original amount}}$$

Step 3. Change $\frac{1}{4}$ to a decimal by dividing 4 into 1. Move the decimal point in the answer two places to the right to find the percent.

Answer: 25%

$$
\begin{array}{r}
.25 \\
4\overline{)1.00} \\
8 \\
\hline
20 \\
20
\end{array}
= 25\%
$$

Solve each percent increase or percent decrease problem below.

Percent Increase

$$\% \text{ increase} = \frac{P \text{ (amount of increase)}}{W \text{ (original amount)}}$$

1. During last year, the price of a certain compact car rose from $7,500 to $7,800. What percent increase in price is this?

2. When Jill started GED class in September, there were 25 students enrolled. By January the class had grown to 30 students. By what percent did Jill's class grow?

3. In response to a rise in cheese prices, Tino's Pizza raised the price of its large pizza from $10.00 to $11.50. What percent price increase is this?

Percent Decrease

$$\% \text{ decrease} = \frac{P \text{ (amount of decrease)}}{W \text{ (original amount)}}$$

4. The value of Arnold's new $10,000 pickup depreciated to a value of $8,000 by the end of the first year he owned it. What percent decrease in value is this?

5. In order to attract more business, Bob's Hamburgers lowered the price of its cheeseburger from $2.00 to $1.75. What percent price reduction is this?

6. On her diet, Jesse's weight dropped to 135 pounds from 150 pounds. What percent of her original body weight did Jesse lose on this diet?

Mixed Practice

7. Marvin was given a raise last month. His monthly salary went from $1,250 to $1,312.50. What percent of raise is this?

8. Before cooking the ham, Martha measured its weight to be 16 pounds. After cooking, the ham weighed 14 pounds. How much "shrinkage" occurred during cooking? (Shrinkage is percent decrease in weight due to water and oil loss.)

9. What percent of a foot is 4 inches? (Hint: The answer will contain a fraction.)

10. On an unpaid balance of $46 on her credit card, Francis was charged $.92 during the month of May. Figure out the monthly interest rate (percentage increase) that Francis is paying for the use of this card.

Finding the Whole When a Part Is Given

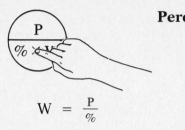

$W = \frac{P}{\%}$

Percent Sentence: To find a whole when part of it is given, divide the part by the percent.

How to do it: Change the percent to either a decimal or fraction and then divide.

Example: 40% of what number is 32? (or, "32 is 40% of what whole number?")

Method 1	Method 2
Step 1. Change 40% to a decimal. 40% = .40	*Step 1.* Change 40% to a fraction. $40\% = \frac{40}{100} = \frac{4}{10} = \frac{2}{5}$
Step 2. Divide .40 into 32. Divide by .4 since .40 = .4.	*Step 2.* Divide 32 by $\frac{2}{5}$.

Method 1:

$$\begin{array}{r} 8\,0. \\ 4\,)\overline{32.0} \\ \underline{32} \\ 0 \\ \underline{0} \end{array}$$

Answer: 80

Method 2:

$$\frac{32}{1} \div \frac{2}{5} = \frac{32}{1} \times \frac{5}{2}$$
$$= \frac{^{16}\cancel{32}}{1} \times \frac{5}{\cancel{2}_1} = 80$$

Answer: 80

Using either method, solve the following problems.

1. 50% of what number is 19?

2. 75% of what number is 27?

3. 11 is 25% of what number?

4. 14 is 7% of what number?

5. $33\frac{1}{3}\%$ of what number is 123? (Use the fraction form of $33\frac{1}{3}\%$.)

6. 56.8 is 8% of what number?

Word Problems: Finding a Whole When a Part Is Given

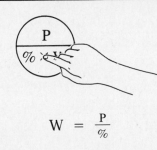

$$W = \frac{P}{\%}$$

In each problem below, remember to divide the given amount (the part) by the percent. To divide, you can change the percent to either a decimal or a fraction.

When the percent is less than 100%, your answer (the whole) will be larger than the part you start with.

1. Fred pays 25% of his monthly salary for rent. If his monthly rent is $245, what is his monthly salary?

2. When she bought a used Ford, Kelly made a down payment of $480. This amounted to 15% of the purchase price. How much did Kelly pay for the Ford?

3. During May's special election, only 34% of the registered voters went to the ballot box. If 14,280 votes were cast, how many registered voters live in the district?

4. To pass the English test, Cillia needs to get 52 answers correct. If 52 questions is 65% of the test, how many questions are on the test?

5. Ahmad was charged $27 during August on the unpaid balance on his Great Western credit card. If Great Western charges a finance charge of 1.5% each month on the unpaid balance, what was Ahmad's unpaid balance at the beginning of August?

6. During the last 3 months, Sal lost 14 pounds while dieting. If this weight is 8% of his original weight, what was Sal's weight before he started the diet?

7. Maria bought a skirt on sale for $28. This is 80% of the original price. Knowing this, figure out the original price of the skirt.

Finding the Original Price

A common two-step problem involves finding an original price when you know the sale price and you know the discount rate. In this problem, the percent circle is used in the second step of the solution.

Example: Jeana bought a dress that was marked "30% off." What was the original price of the dress if Jeana paid only $49?

Step 1. The first step is to determine what percent $49 is of the original price. To do this, subtract 30% from 100%.
100% − 30% = 70%   **$49 is 70% of the original price**

Step 2. Now ask yourself, "70% of what number is $49?" Solve this problem as you solved the problems on the previous page. Divide $49 by 70% in either decimal or fraction form.
70% = .70 = .7

$$.7 \overline{)\$49.0} \quad \begin{array}{c} 70. \end{array}$$

$$W = \frac{P}{\%} = \frac{\$49}{70\%}$$

Answer: $70

---

8. By paying cash, Paul got a 6% discount on a television set. If he paid $451.20 cash for the set, how much would he have paid if he didn't pay cash?

9. Rod bought a suit that was marked "25% off." If he paid $129 for the suit on sale, what was the suit's regular price?

10. If she pays her car payment before the 10th of each month, Ella is given a special 2% discount on the payment. If she pays $140 before the 10th, how much would Ella pay after the 10th?

11. Lolla went to a special Winter Sale where all furniture was marked 35% off. What was the regular price of a recliner that Lolla got on sale for $256?

Percent Word Problems: Mixed Practice

Percent Circle

Use the percent circle to solve each word problem below. As a first step in each problem, decide if you are looking for the part, the percent, and the whole.

Be careful; some problems — such as percent increase and percent decrease — may require two steps to solve.

1. According to the evening newspaper, only 9% of the city's residents ride the bus more than once each month. If 75,400 people live in the city, how many ride the bus twice or more each month?

2. Out of Jeffrey's monthly salary of $840, his employer withholds $67.20 for state income tax. What percent of Jeffrey's salary is withheld for this tax?

3. Benson pays $360 each month in rent. However, if he pays on the 1st of each month, he is given a 3% discount. How much should Benson write his rent check for each month he pays on the 1st?

4. During a snowstorm, only 36% of the employees of Amer Electronics Company were able to get to work. If only 90 employees reported for work, how many people work for this company?

5. Jordan bought a rug on sale for $156. If this price is 75% of the original price, what was the cost of the rug before the sale?

6. Due to an early freeze in Florida last year, the price of orange juice increased this year. The price of Diamond O's 12-ounce size of orange juice concentrate rose from $1.20 to $1.38. What percent increase in price is this?

7. During the month of December, 42% of the babies born at St. Mary's Hospital were girls. If 150 babies were born there during December, how many were girls?

8. Grace bought a sweater at an "After Christmas Sale" that was marked down 30% from its original price. If she paid $21.50 for the sweater on sale, what was its price before the sale?

9. Andrea contacted 185 residents of Wittenville. She asked each person how he or she felt about the upcoming budget election. According to her results as shown at right, what percent of residents were planning to vote against the budget?

Budget Election Survey

For	Against	Undecided
83	74	28

10. Frieda decided that the May sale was a good time to buy a new television set. At Frank's Discount Center she could get a good deal, but she had to make a down payment of 15%. Frieda chose a new set from the sizes listed at right. If she made a down payment of $64.50, which size of set did she buy?

Model	Original Price	Sale Price
14 in.	$249	$189
17 in.	$399	$329
19 in.	$569	$430
22 in.	$749	$619

In problems 11 and 12, circle the arithmetic expression that will give the correct answer to each question. You do not need to solve these two problems.

11. John LaPlace owns 648 acres of farmland, which he uses for growing crops and raising sheep. Early this spring he plans to plant 360 acres in corn. If this corn takes up 72% of the total acreage he uses for crops, on how many acres does John plant crops?

a) $648 \div .72$
b) $(648 - 360) \div .72$
c) $360 \div .72$
d) $.72 \times (648 - 360)$
e) $.28 \times (648 - 360)$

12. After raising the price of the large-size cola, Cinema Center has a 25% decrease in the number of large-size colas sold. Before the price increase, they could sell about 340 large-size colas at each movie. About how many can they expect to sell now, after the price increase?

a) $340 - .25 \times 340$
b) $340 + .25 \times 340$
c) $.25 \times 340 - 340$
d) $.25 \times 340$
e) $340 - .25 + 340$

Percent Skills Review

To complete the chapter on percent, check your percent computation skills on this page. Check each answer carefully.

Changing Percents to Decimals and Fractions:
Review pages 116 to 120.
Change each percent below to both a decimal and a fraction. Write the decimal answer on the first line and the fraction answer on the second line.

1. 50% _____ _____   25% _____ _____   60% _____ _____

2. 75% _____ _____   12.5% _____ _____   2.5% _____ _____

Finding Part of a Whole: Review pages 126 to 127.
Determine each number indicated below.

3. 40% of 60   25% of 200   8% of 12 $33\frac{1}{3}$% of 96

4. .8% of 14   5.5% of 25 $\frac{4}{5}$% of 100 $\frac{3}{4}$% of 50

5. 8.8% of $5,000   300% of 100   125% of 42   250% of 18

Finding What Percent a Part Is of a Whole:
Review page 132.

6. 8 is what percent of 32?   7. 12 is what percent of 96?

8. What percent of 200 is 31?   9. What percent of 42 is 14?

10. 150 is what percent of 75?   11. What percent of 30 is 66?

12. $33\frac{1}{3}$ is what percent of 100?   13. $\frac{1}{2}$ is what percent of 100?

Finding a Whole When Part of It Is Given:
Review page 136.

14. 25% of what number is 19?   15. 50% of what number is 31?

16. 51 is 20% of what number?   17. .3% of what number is 99?

18. $66\frac{2}{3}$% of what number is 32?   19. $33\frac{1}{3}$% of what number is $13\frac{1}{3}$?

20. 12% of what amount is $27.06?   21. $12.76 is 8.8% of what amount?

6
Special Topics in Math

TOPIC 1: Using Approximation in Problem Solving

Until now you have worked each problem to get an exact answer. We want to look at a shortcut called **approximation**.

An approximation (or approximate answer) is a number that is "about equal" to an exact answer. Being able to find an approximation is useful in three ways:

- to find an approximate answer when an exact answer is not needed

- to quickly check your results after you work a problem exactly

- to quickly pick the most reasonable answer from among answer choices given on some test questions

Approximation is very useful in problems involving mixed decimals, mixed numbers, and percents. To find an approximation, you replace one or more numbers in a problem with whole numbers. **You choose a whole number that is as close as possible to the mixed decimal or mixed number it replaces.**

Replacing a mixed decimal or mixed number with a whole number is called **rounding to the nearest whole number.** Rounding can be used to quickly find approximations in multiplication and division problems.

Example 1: Find an approximation for the product of $4\frac{7}{8}$ times $7\frac{1}{8}$.

Approximation $5 \times 7 = 35$

Exact Answer
$$4\frac{7}{8} \times 7\frac{1}{8}$$
$$= \frac{39}{8} \times \frac{57}{8}$$
$$= \frac{2,223}{64}$$
$$= 34\frac{47}{64}$$

Step 1. Round each mixed number to the nearest whole number.

$$4\frac{7}{8} \to 5 \qquad 7\frac{1}{8} \to 7$$

Step 2. Multiply the whole numbers.

Answer: 35

Note: 35 is close to the exact answer $34\frac{47}{64}$, and it is much easier to compute!

Example 2: Find an approximation for the quotient of 322.344 divided by 3.96.

Step 1. Round each mixed decimal to the nearest whole number.

322.344 → 322

3.96 → 4

Step 2. Divide the whole numbers.

Answer: 80.5

Approximation

```
      80.5
 4) 322.0
    32
     2 0
     2 0
        0
```

Exact Answer

```
            81.4
 3.96 ) 322.34.4
        316 8
          5 54
          3 96
          1 58 4
          1 58 4
                0
```

Write an approximation for each number below. The first problem in each row is done as an example.

1. $2\frac{1}{6}$ _2_ $12\frac{2}{9}$ _____ $8\frac{1}{8}$ _____ $5\frac{3}{16}$ _____

2. 12.89 _13_   5.03 _____   14.975 _____   2.025 _____

3. 8.8% _9%_   4.95% _____ $5\frac{9}{10}$% _____ $13\frac{1}{8}$% _____

Compute an approximate answer for each problem below by first rounding each number to the nearest whole number. The first problem in each row is done as an example. Do *not* work out the exact answer.

Addition and Subtraction: Approximation is most useful in mixed number problems.

4.
$$5\frac{7}{8} \to 6$$
$$\underline{+3\frac{1}{16} \to +3}$$
$$9$$

$$13\frac{11}{12}$$
$$\underline{+\ 7\frac{5}{6}}$$

$$4\frac{8}{9}$$
$$\underline{+2\frac{1}{5}}$$

$$7\frac{15}{16}$$
$$\underline{+4\frac{1}{6}}$$

5.
$$6.3 \to 6$$
$$\underline{-2.9\quad -3}$$
$$3$$

$$21.9$$
$$\underline{-13.2}$$

$$9.1$$
$$\underline{-5.9}$$

$$13.8$$
$$\underline{-\ 8.1}$$

Multiplication and Division: Approximation is useful in all problems.

6. $4\frac{1}{8} \times 3\frac{7}{8}$

$4 \times 4 = 16$

$5\frac{9}{10} \times 8\frac{1}{5}$

$11\frac{7}{8} \times 4\frac{1}{7}$

$3\frac{1}{3} \times 5\frac{8}{9}$

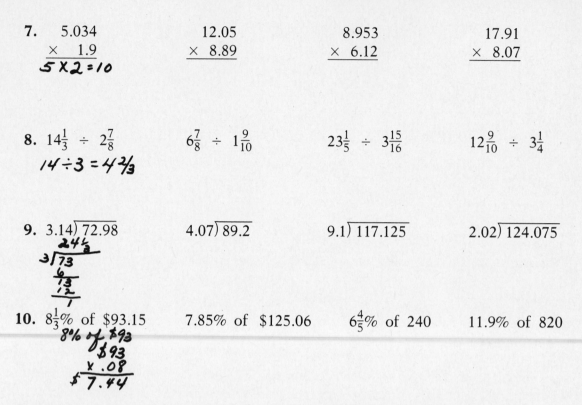

7.
5.034
× 1.9
5×2=10

12.05
× 8.89

8.953
× 6.12

17.91
× 8.07

8. $14\frac{1}{3} \div 2\frac{7}{8}$
14÷3=4⅔

$6\frac{7}{8} \div 1\frac{9}{10}$

$23\frac{1}{5} \div 3\frac{15}{16}$

$12\frac{9}{10} \div 3\frac{1}{4}$

9. $3.14\overline{)72.98}$
24⅓
3)73
6
13
12
1

$4.07\overline{)89.2}$

$9.1\overline{)117.125}$

$2.02\overline{)124.075}$

10. $8\frac{1}{3}\%$ of \$93.15
8% of $93
$93
× .08
$7.44

7.85% of \$125.06

$6\frac{4}{5}\%$ of 240

11.9% of 820

Compute an *approximate answer* for each problem below. Then, using this approximation as a clue, circle the *exact answer* from the answer choices given.

11. $5\frac{6}{7} \times 8\frac{1}{9}$
  a) $38\frac{19}{63}$
  b) $47\frac{32}{63}$
  c) $54\frac{57}{63}$

12. $24\frac{7}{8} \div 2\frac{3}{4}$
  a) $4\frac{7}{22}$
  b) $6\frac{3}{22}$
  c) $9\frac{1}{22}$

13. $3\frac{6}{7} \times 5$
  a) $19\frac{2}{7}$
  b) $24\frac{1}{7}$
  c) $27\frac{6}{7}$

14.
5.078
× 2.96
  a) 12.90538
  b) 15.03088
  c) 19.09828
  d) 23.15948

15. $2.9\overline{)14.674}$
  a) 3.76
  b) 4.06
  c) 4.51
  d) 5.06

16. 5.15% of 19.8
  a) 1.01970
  b) 1.43640
  c) 1.95036
  d) 2.01634

17. 24.9% of \$145.87
  a) \$21.62
  b) \$27.83
  c) \$36.32
  d) \$42.93

Using Approximation in Word Problems

Approximation is an especially useful tool for solving word problems.

- You can use approximation to help you decide which operation is correct for the problem you're solving.

- You can use approximation to check the accuracy of your exact answer.

Example: Last Saturday, Peggy worked 5.75 hours of overtime. If her overtime salary rate is $8.16 per hour, how much overtime pay did she earn that day?

An approximate answer can be quickly computed by replacing each number with a whole number and then multiplying.

$5\frac{3}{4} \to 6$ $\$8.16 \to \$8$

Approximate answer: $6 \times \$8 = \mathbf{\$48}$

The exact answer is $46.92.

Exact Answer
```
   $8.16
 × 5.75
 40 80
 571 2
 4080
$4692 00
=$46.92
```

Notice how the approximate answer $48 serves as a useful clue:

- First, the amount $48 seems about right. This tells us that multiplication is the correct operation to use.

- Second, you know that the exact answer must be close to $48. If you compute an exact answer that is several dollars away from $48, you'll know that you probably made a mistake. You would then want to recheck your multiplication.

In each of the following problems, compute both an approximate answer and an exact answer. Write the approximation on the first line and the exact answer on the second line. The first problem is completed as an example.

1. If chicken is on sale for $1.04 per pound, how much would you pay for a whole chicken that weighs 5.95 pounds?

   approx.
   $1.04 → $1.00
   5.95 → × 6
   $6.00

   Exact
   ```
     $1.04
   × 5.95
     520
     936
     520
   $6.1880
   ```
   $6.19 (to nearest penny)

   $6.00 underline approx. $6.19 underline exact

2. How much rock can Bryce carry with his truck in 8 loads if he can carry $2\frac{7}{8}$ tons on each load?

_____ _____
  approx.   exact

3. Tabatha bought a package of hamburger that cost $5.89. If the package weighs 2.08 pounds, how much is Tabatha paying per pound?

_____ _____
  approx.   exact

4. At birth Sunny was $20\frac{1}{8}$ inches long. By her 1st birthday she had grown $9\frac{1}{16}$ inches. By her 2nd birthday she had grown another $4\frac{3}{4}$ inches. Use this information to determine Sunny's height on her 2nd birthday.

_____ _____
  approx.   exact

5. If 197 of the 503 parents attended the school Christmas play, what percent of parents showed up? Round your exact answer to the nearest percent.

_____ _____
  approx.   exact

6. What is the difference in width of a shelf that is $19\frac{1}{8}$ inches wide and one that is $11\frac{15}{16}$ inches wide?

_____ _____
  approx.   exact

7. Jeremy lives in a state that has a 6.9% sales tax. By adding on the sales tax, what would Jeremy pay for a steam iron that has a price tag of $34.19?

_____ _____
  approx.   exact

8. At a weight of $8\frac{1}{3}$ pounds per gallon, what is the weight of $35\frac{5}{6}$ gallons of water?

_____ _____
  approx.   exact

9. How many concrete blocks $10\frac{7}{8}$ inches long will it take to cross a patio that measures 33 feet wide? Assume that the blocks are laid end-to-end. (Hint: As a first step, change the patio width to inches only.)

_____ _____
  approx.   exact

Use approximation to help you quickly choose the correct answer for each problem below. Do not work out the exact answer.

10. If salmon steak is on sale for $3.98 per pound, how much would a package of steaks that weighs 5.037 cost?

a) $14.95
b) $17.26
c) $20.05
d) $24.76
e) $27.89

11. Three hoses are joined together end-to-end. Determine the length of the combined hoses if the separate hose lengths are as follows: $26\frac{5}{6}$ feet, $31\frac{11}{12}$ feet, and $19\frac{1}{8}$ feet.

a) $71\frac{7}{8}$ feet

b) $73\frac{11}{12}$ feet

c) $75\frac{3}{4}$ feet

d) $77\frac{7}{8}$ feet

e) $98\frac{8}{9}$ feet

12. What is the difference in length between a shaft that measures 24.0008 inches and a shaft that measures 17.0909 inches?

a) 5.1099 inches
b) 5.8009 inches
c) 6.0999 inches
d) 6.9099 inches
e) 7.9909 inches

13. At a price of 1.23\frac{9}{10}$ per gallon, how much would you pay for $14\frac{1}{10}$ gallons of unleaded gasoline?

a) $17.47
b) $18.69
c) $19.82
d) $21.41
e) $23.08

14. Ann's roof measures 60 feet across. If each roof shingle is $5\frac{3}{4}$ inches wide, about how many shingles placed side-by-side would it take to cross this roof?

a) 105
b) 126
c) 150
d) 200
e) 230

15. At a clearance sale, the $39.97 original price of a dress was marked down to $15.99. About what percent decrease in price is this reduction?

a) 43%
b) 50%
c) 60%
d) 71%
e) 79%

16. During the year, the price of a new dishwasher rose by 8.9%. If last year's price was $297.89, what can you expect to pay this year?

a) $324.40
b) $382.60
c) $401.32
d) $417.95
e) $459.00

Learning That Approximation Doesn't Always Help

On the previous few pages you saw how useful approximation can be. On this page we want to remind you that approximation is only a useful tool. **You should not rely on approximation to take the place of exact work.**

As you've seen, **approximation works well when answer choices differ greatly in value from each other.** In these problems, your approximate answer clearly helps you choose the correct answer. However, when answer choices are close in value, approximation may lead you to choose a wrong answer! There are two types of problems where you must be careful:

When the numbers are smaller than 1. For example, to divide $\frac{5}{8}$ by $\frac{15}{16}$, it doesn't help to find an approximate answer by replacing both fractions by the number 1. The approximate answer is also 1. An approximate answer of 1 won't help you choose from among the answer choices shown at right.

Answer Choices

a) $\frac{9}{16}$ c) $\frac{2}{3}$

b) $\frac{11}{27}$ d) $\frac{7}{9}$

When the answer choices are very close in value. For example, to multiply 3.92 by 2.03, you might write an approximate answer of 8 (found by multiplying 4 by 2).

As you look among the answer choices, you see that each is so close to 8 that you can't be sure which is correct. Incidentally, answer *c* is not correct!

a) 7.8366
b) 7.9576
c) 8.0046
d) 8.1456
e) 8.2046

Use approximation to try to solve each problem below. Then, in your own words, explain why approximation doesn't work in each problem. (The correct answer to each problem is indicated.)

1. Shannon bought a 7.08- pound roast that was on sale for $1.93 per pound. How much did this roast cost her?

   a) $13.46
  *b) $13.66
   c) $13.86
   d) $14.16

2. What is $\frac{6}{7}$ of $\frac{15}{16}$ of an inch?

  *a) $\frac{45}{56}$ of an inch
   b) $\frac{79}{112}$ of an inch
   c) $\frac{7}{8}$ of an inch

3. If only 98 out of 202 employees showed up for work during a snowstorm, what percent of employees made it in to work?

   a) 45.3%
   b) 47.2%
   c) 47.9%
  *d) 48.5%

TOPIC 2: Measurement
Measuring with an English Ruler

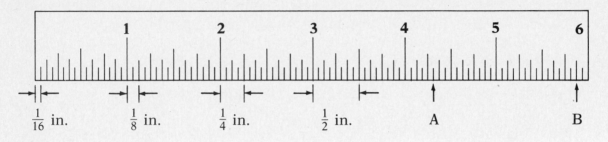

Pictured above is a six-inch English ruler. One-inch units are numbered 1 to 6. Each inch is divided into half inches (every 8 lines), quarter inches (every 4 lines), eighth inches (every 2 lines), and sixteenth inches (from one line to the next). Examples of fractions of an inch are shown above.

You read a ruler from left to right. To make reading a ruler easier, each fraction of an inch is represented by a line of different height. For example, the line at each half inch is higher than the line at each quarter inch. The line at each quarter inch is higher than the line at each eighth inch, and so on.

When measuring, always reduce fraction answers when possible.

Example 1: How far is point A from the left end of the ruler?

First notice that point A is between 4 and 5 inches from the left end.

Because point A is at a $\frac{1}{16}$-inch line, count the number of sixteenths that point A is beyond 4 inches. Point A is 5 sixteenths to the right of the 4-inch line.

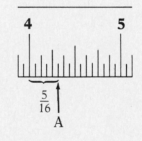

Answer: Point A is $4\frac{5}{16}$ **inches** from the left end of the ruler.

Example 2: How far is point B from point A?

Step 1. Write down the distance that each point is from the left end of the ruler.

Point A is $4\frac{5}{16}$ inches from the left end.

Point B is $5\frac{7}{8}$ inches from the left end.

Step 2. To find how far point B is from point A, subtract the two distances found in step 1.

$$5\frac{7}{8} = 5\frac{14}{16}$$
$$-4\frac{5}{16} = -4\frac{5}{16}$$
$$\overline{\qquad 1\frac{9}{16}}$$

Answer: Point B is $1\frac{9}{16}$ **inches** from point A.

Answer each of the following questions about the English ruler shown on page 149.

1. What are the smallest parts into which each inch is divided?

2. How many quarter inches is each inch divided into?

3. What is the distance between each pair of numbers given on the ruler?

4. How far is point B from the left end of the ruler?

Answer the following questions about the ruler shown below.

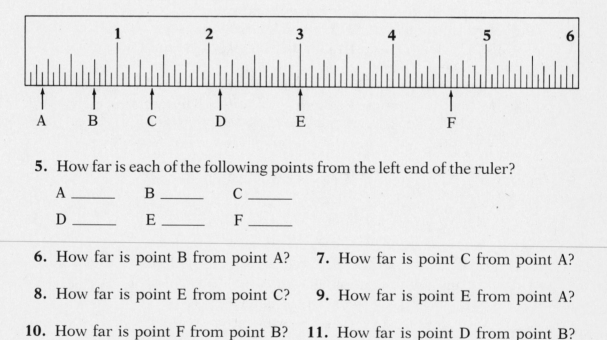

5. How far is each of the following points from the left end of the ruler?

A _____   B _____   C _____

D _____   E _____   F _____

6. How far is point B from point A?   7. How far is point C from point A?

8. How far is point E from point C?   9. How far is point E from point A?

10. How far is point F from point B?   11. How far is point D from point B?

Measuring with a Metric Ruler

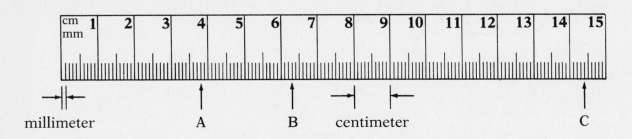

Pictured above is a 15-centimeter metric ruler. One-centimeter units are numbered 1 to 15. Each centimeter is divided into 10 millimeters.

A centimeter ruler is much easier to read than an English ruler. On a centimeter ruler no fractions are used. Instead, you read a distance as a number of centimeters (cm) plus a number of millimeters (mm). For example, point B is 6 cm 3 mm from the left end of the ruler.

Because there are 10 mm in 1 cm, each millimeter can be written as .1 centimeter. For example, 3 mm = .3 cm. Therefore, 6 cm and 3 mm can be written as 6.3 cm.

To write a distance in centimeters only, write millimeters as the first number to the right of the decimal point.

Example 1: Write 4 cm 9 mm as centimeters only.

  Answer: 4 cm 9 mm = 4.9 cm

Example 2: Write 7.5 cm as a number of centimeters and millimeters.

  Answer: 7.5 cm = 7 cm 5 mm

To find the distance between two points on a centimeter ruler, write each distance as a decimal number and then subtract.

Example 3: On the ruler above, how far is point B from point A?

  Step 1. Using decimal numbers, write the distance that each point is from the left end of the ruler.
  Point B = 6 cm 3 mm = 6.3 cm
  Point A = 3 cm 8 mm = 3.8 cm

  Step 2. Subtract the distances.

$$\begin{array}{r} 6.3 \text{ cm} \\ -3.8 \text{ cm} \\ \hline 2.5 \text{ cm} \end{array}$$

  Answer: 2.5 cm or 2 cm 5 mm

Answer each of the following questions about the metric ruler shown above.

1. What are the smallest divisions of the metric ruler called?

2. What are the numbered divisions called?

3. How many millimeters are there in 1 centimeter?

4. How far is point C from the left end of the ruler?

Write each of the following distances in the form indicated.

5. 3 cm 4 mm = _____ cm   7 cm 1 mm = _____ cm

6. 2.6 cm = __ cm __ mm   8.3 cm = __ cm __ mm

Answer the following questions about the ruler shown below.

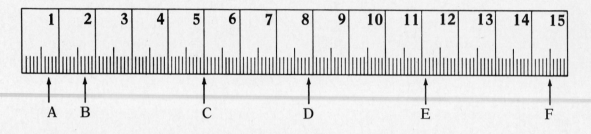

7. How far is each of the following points from the left end of the ruler?

A _____   B _____   C _____

D _____   E _____   F _____

8. How far is point B from point A?

9. How far is point C from point A? (express your answer in cm only)

10. How far is point F from point C? (express your answer as a number of centimeters and millimeters)

11. How far is point D from point F? (express your answer two ways)

152

TOPIC 3: Understanding Simple Interest

Interest is money that is earned (or paid) for the use of money.

- If you deposit money in a savings account, interest is money that the bank pays you for using your money.

- If you borrow money, interest is money that you pay for using the lender's money.

Interest is earned (or paid) on **principal** — the amount that is deposited or borrowed. **Simple interest** can be found if the original principal does not change. We want to study simple interest for several reasons:

1. Interest is part of both the saving and the borrowing of money.

2. Interest problems involve the use of whole numbers, decimals, fractions, and percents. Thus, interest problems give you a chance to practice several math skills at one time.

3. To solve an interest problem, we use the **simple interest formula**. A formula is a rule that uses letters to stand for words. Learning to use formulas is an important part of your study of mathematics.

The Simple Interest Formula

To determine an amount of interest, you **multiply the principal by the rate by the time**.

In words, Interest = Principal × Rate × Time

As a formula, $\boxed{\text{I} = \text{PRT}}$

where, I = interest, written in dollars
P = principal: money deposited or borrowed, written in dollars
R = percent rate, written as a fraction or a decimal
T = time, written in years or parts of a year

Written in symbols, I = PRT is called the **simple interest formula**. This formula is used to compute both the interest earned and the interest paid.

Note: Be sure not to confuse the use of the letter P in the simple interest formula with its use in the percent circle. In I = PRT, P stands for *principal*. In the percent circle, P stands for *part*.

Example 1: What is the interest earned on $400 deposited for 2 years in a savings account that pays 6% simple interest?

Step 1. Identify P, R, and T.

P = principal, the amount deposited = $400

R = percent rate of interest = 6% = $\frac{6}{100}$

T = time, the number of years the money is deposited = 2

In this problem, we write the percent rate as a fraction because we see that canceling can be used to simplify the multiplication. In other problems, multiplication may be easier if you write the percent as a decimal.

Step 2. To compute the interest, first replace the letters in the formula with the values identified in step 1. Then multiply.

$$I = PRT = 400 \times \frac{6}{100} \times 2$$
$$= \overset{4}{\cancel{400}} \times \frac{6}{\cancel{100}_1} \times 2 = 48$$

Answer: $48

In Example 2, we can write the percent either as a decimal or as a fraction. We choose to write it as a fraction because we can use cancellation to simplify the computations. As a shortcut, we write the fraction as a mixed number $(6\frac{1}{2})$ over 100. As you see, this is not confusing because the denominator (100) cancels out in the first multiplication done in step 2.

Example 2: What is the interest paid on a $1,400 loan borrowed for 4 years if the interest rate is 6½%?

Step 1. Identify P, R, and T.

P = principal, the amount of loan = $1,400

R = percent rate of interest = 6½% = $\frac{6\frac{1}{2}}{100}$

T = time, the number of years the money has been borrowed for = 4

Step 2. To compute the interest, first replace the letters in the formula I = PRT. Then multiply.

$$I = PRT = 1400 \times \frac{6\frac{1}{2}}{100} \times 4$$
$$= \overset{14}{\cancel{1400}} \times \frac{6\frac{1}{2}}{\cancel{100}_1} \times 4$$
$$= 14 \times \frac{13}{\cancel{2}_1} \times \overset{2}{\cancel{4}} \quad (\text{since } 6\frac{1}{2} = \frac{13}{2})$$
$$= 14 \times 13 \times 2$$
$$= \$364$$

Answer: $364

In the problems below, use the simple interest formula I = PRT to find either the interest earned or the interest paid.

Interest Earned

1. What is the interest earned on $500 deposited for 3 years in a savings account that pays 5% simple interest?

2. How much interest would Ms. Phipps earn on a deposit of $1,500 in 1 year if the interest rate was $5\frac{1}{2}$% per year?

Interest Paid

3. At an 11.5% interest rate, how much interest would Sharon have to pay on an $850 loan borrowed for 3 years?

4. How much interest would Andy pay for a loan of $1,200 at 12% if he repaid it at the end of 2 years?

Finding the Total Balance or the Total Owed

Many problems ask you to find the total balance or the total owed at the end of a time period. In these problems, you add the interest to the principal in order to find the total.

Example: Mark deposited $250 in a savings account that earns 5.5%. What will be the total in Mark's account after 2 years?

Step 1. Compute the interest earned in 2 years.

$$I = PRT$$

			250	P	13.75	PR
	P = $250		×.055	× R	× 2	× T
	R = 5.5% = .055		1 250	PR	27.50	PRT
	T = 2		12 50			
			13.750			

$$I = \$27.50$$

Step 2. Add the $27.50 interest to the principal $250.

$$\begin{array}{r} \$250.00 \\ +\quad 27.50 \\ \hline \$277.50 \end{array}$$

Answer: $277.50

Solve each problem below.

5. Earning simple interest of 5%, what total will be in a savings account of $750 left for a period of 3 years?

6. What will be the total amount owed on a loan if the principal is $2,000, the interest rate is 15%, and the time is 2 years?

7. How much is now in a savings account if the amount deposited was $275, the interest rate was 6%, and the time was 4 years?

8. What amount will Luanne have to repay the bank after 2 years for a loan of $1,500 borrowed at $9\frac{1}{2}$% simple interest?

Interest for Part of a Year

Although interest is earned (or paid) as a yearly interest rate, not all deposits or loans are made for whole years. Some are made for a part of a year. To use the simple interest formula for part of a year, write the time either as a decimal or as a fraction.

In example 1, both the interest rate and the time are written as fractions, and canceling is used to simplify the multiplication.

Example 1: What is the interest earned on $300 deposited for 9 months at an interest rate of 6%?

Step 1. Identify P, R, and T.
$$P = 300 \qquad R = 6\% = \tfrac{6}{100} \qquad T = 9\,\text{months} = \tfrac{9}{12} = \tfrac{3}{4}\,\text{year}$$

Step 2. Replace the letters in the formula $I = PRT$ with the values identified in step 1, and then multiply.

$$I = PRT = 300 \times \tfrac{6}{100} \times \tfrac{3}{4}$$
$$= {}^{3}\cancel{300} \times \tfrac{{}^{3}\cancel{6}}{\cancel{100}_{1}} \times \tfrac{3}{\cancel{4}_{2}} = \tfrac{27}{2} = 13\tfrac{1}{2} = 13.50$$

Note: We write $\tfrac{1}{2}$ as .50 because we want the answer to be in dollars and cents.

Answer: $13.50

In example 2, the time is given as a mixed number. In this problem it looks easier to write time as a mixed decimal and write R as a decimal and then multiply.

Example 2: How much interest is paid on a loan of $525 borrowed for $3\tfrac{1}{2}$ years at a 9% simple interest rate?
$$P = 525 \qquad R = 9\% = .09 \qquad T = 3\tfrac{1}{2} = 3.5$$

$$I = PRT = 525 \times .09 \times 3.5$$

```
  525          47.25
× .09         ×  3.5
47.25        236 25
             141 75
             165.375   = 165.38 Rounding to the nearest cent.
```

Answer: $165.38

156

In the following problems, decide whether you are looking for the interest only or the new total balance. Then solve each problem using either decimals or fractions.

1. Olsen Furniture charges 13% simple interest on all purchases. How much interest would Heather pay on a purchase of $750 if she paid the full amount at the end of 6 months?

2. Brady deposited $350 in a savings account at his bank. How much interest will his money earn after 15 months if he is paid an interest rate of 6%?

3. To buy a used car, Chris signed a note with the dealer agreeing to pay the $2,500 balance in 18 months. The dealer charged 11% interest for the financing. What total did Chris agree to pay the dealer at the end of the time period?

4. Borrowing at a 14% interest rate, how much interest would you pay on a loan of $5,000 held for 5 years?

5. Figure out the interest earned on $650 deposited for 2 years and 6 months in a savings account that pays 7% interest.

6. How much total money will Jennifer have to repay the bank if she borrows $900 for 9 months at an interest rate of $11\frac{1}{2}$%?

7. Super Stereo charged Myrna 18% interest on her $650 purchase. How much interest must Myrna pay to Super Stereo if she pays the whole amount at the end of 10 months?

8. Amber and her husband Don borrowed $1,400 from their credit union. They signed a note saying that they would pay 8% interest and would pay back the entire amount at the end of 7 months. How much would be due at the end of that time?

Simple Interest Problems: Mixed Practice

On this page you'll have a chance to review several types of simple interest problems.

Questions 1 through 4 refer to the following story.

Mrs. James opened a savings account at her bank. She deposited $2,800 in an account that paid $6\frac{1}{2}\%$ simple interest. At the end of 18 months, Mrs. James withdrew all her money. She wanted to use this money as a down payment on a car.

The car she decided to buy was a two-year-old Honda. The asking price was $7,000. After Mrs. James got the dealer to reduce the price by 10%, she agreed to buy the car. She gave the dealer all the money she had withdrawn from the bank. She agreed to pay off the balance in one payment at the end of 9 months. For this 9-month loan, the dealer charged her an 11% yearly interest rate.

1. How much money did Mrs. James withdraw from her savings account? (Include both principal and interest.)

2. What price did Mrs. James agree to pay for the car?

3. After giving the dealer her money as a down payment, how much did Mrs. James still owe for the car?

4. At the end of 9 months, how much did Mrs. James have to pay the dealer in order to pay off the loan?

Questions 5 and 6 refer to the story below.

Kevin applied for and received a charge card at Shopper's Department Store. When signing for the card, he agreed to pay the store 18% yearly interest on the card's balance. The 18% would be actually charged as a 1.5% finance charge on any unpaid balance that was 1 month overdue.

Before he left the store, Kevin decided to buy a table saw that was on sale. The original price of the saw was $420. On it was a sales tag that read "25% off the original price." Kevin bought the saw at the sales price and charged it on his card.

5. What price was Kevin charged for the table saw?

6. When the balance became 1 month overdue, what total amount did Kevin owe? (Include both principal and interest.)

TOPIC 4:
Introduction to Data Analysis

Before we define *data analysis*, you first need to know what we mean by *numerical data*.

> **Numerical data** are a group (often called **set**) of numbers that are related in some way.

Here is an example of numerical data:

- the set of numbers that stand for the ages of all the students in your class.

| 21 years |
| 20 years |
| 42 years |
| 18 years |

Data analysis is the study of numerical data. Data analysis is used in decision making at all levels of society. It is used in schools, businesses, and government agencies. You use data analysis yourself when you count the change in your pocket to see how much you can spend for lunch!

The Language of Data Analysis

Three words that are often used in data analysis are **mean**, **median**, and **ratio**. We'll look at the definition and use of each word on the pages ahead.

MEAN

> **Mean** is another word for **average**. You may already know the two steps used to find the average of a set of numbers:

1. First, compute the sum of the set.

2. Second, divide this sum by the number of numbers in the set.

Example: Find the mean (average) of the following set of numbers:
38, 52, 25, 19, 67, 63

Step 1. Compute the sum of the set.

$$
\begin{array}{r}
38 \\
52 \\
25 \\
19 \\
67 \\
+63 \\
\hline
264
\end{array}
$$

Step 2. Divide this sum by 6, the number of numbers in the set.

$$
\begin{array}{r}
44 \\
6\overline{)264} \\
\underline{24} \\
24 \\
\underline{24} \\
0
\end{array}
$$

Answer: Mean = 44

Compute the average of each set of numbers below.

1. 4, 7, 9, 13, 17   **2.** $2.39, $3.49, $3.99   **3.** $1\frac{1}{2}, 3\frac{3}{4}, 5\frac{1}{4}, 6\frac{1}{2}$

MEDIAN

The ***median*** of a set of numbers is the number that is the middle value. To find the median, follow these two steps:

1. Arrange the numbers in order, from smallest to largest.

2. Count the number of numbers in the set:

    a) For an **odd** number of numbers, the median is the middle number.

    b) For an **even** number of numbers, the median is the average of the two middle numbers.

Example 1: Find the median in the following set of numbers:
9, 13, 4, 7, 17

    Step 1. Arrange the numbers in order, smallest one first.

    Step 2. Since there is an odd number (5) of numbers, the median is the middle value (9).

                      4
                      7
                      9 ← middle number = median
                    13
                    17

Answer: Median = 9

Example 2: What is the median in the following set of numbers?
38, 52, 25, 19, 67, 63

    Step 1. Arrange the numbers in order, smallest one first.

            19, 25, 38, 52, 63, 67

            median is average of two middle numbers

    Step 2. Since there is an even number (6) of numbers, the median is the average of the two middle numbers, 38 and 52.

            38   45 ← median
          +52 2)90
          90   8
                  10
                  10

Answer: Median = 45

Find the median for each set of numerical data below.

4. 6 lbs. 4 oz.
8 lbs. 3 oz.
9 lbs. 15 oz.
6 lbs. 2 oz.
7 lbs. 5 oz.

5. 87°, 92°, 96°, 95°, 93°, 91°, 90°

6. 1.26, $1.32, $1.18, $1.19, $1.20, $1.31

RATIO

A *ratio* is the comparison of two numbers. For example, if there are 8 women in your class and 5 men, the ratio of **women to men is 8 to 5**.

The ratio 8 to 5 can be written in symbols in two ways.

- With a colon, the ratio 8 to 5 is written 8:5.

- As a fraction, the ratio 8 to 5 is written $\frac{8}{5}$.

In words, a ratio is always read with the word *to*. Both 8:5 and $\frac{8}{5}$ are read as ratios as "8 to 5."

A ratio, like a fraction, is usually reduced to lowest terms. However, an improper fraction ratio such as $\frac{8}{5}$ is not changed to a mixed number.

When you write a ratio, you write the numbers in the same order as asked for in the question. Although the ratio of women to men is 8 to 5, the ratio of **men to women is 5 to 8 (or $\frac{5}{8}$)**.

Here is another example of a ratio problem.

Example: A new car gets 24 miles per gallon during city driving and 38 miles per gallon during highway driving. What is the ratio of highway mileage to city mileage?

   Step 1.   Write highway mileage as the numerator of the ratio fraction because it is mentioned first in the question.

$$\frac{38}{24} = \frac{\text{highway mileage}}{\text{city mileage}}$$

   Step 2.   Reduce the ratio fraction $\frac{38}{24}$.

$$\frac{38 \div 2}{24 \div 2} = \frac{19}{12}$$

Answer: 19:12 or $\frac{19}{12}$

In each problem below, remember to express the ratio in lowest terms.

7. On her test, Jan got 26 questions correct out of a total of 32. What is the ratio of her correct answers to the total number of questions?

8. To make pink, Andrea mixes 18 ounces of red paint with 12 ounces of white. In this paint mixture, what is the ratio of white paint to red?

9. Dave makes $7.50 per hour while Sam makes only $4.50. What is the ratio of Dave's salary to Sam's?

10. Find the ratio of 1 foot to 1 yard. (Change 1 yard to 3 feet before writing the ratio as a fraction.)

161

Displaying Numerical Data

Even on tests, numerical data are usually not written just as a group of numbers. Instead, data are presented in the same way you might see them used in magazines and newspapers. As part of this introduction to data analysis, we'll now look at the two most common ways of displaying data: 1) in a **table** and 2) on a **graph**.

Reading a Table

- A *table* is simply an orderly arrangement of numbers.

Numerical data appear in a table as rows and columns of numbers. Rows are read from left to right. Columns are read from top to bottom. Word labels (or symbols) tell what information is contained in each row and column.

Example: Below is a nutrition chart for several types of grain. This chart contains a table of numerical information. Take note of the title of the chart and all the other information included.

Nutrition Chart

Uncooked grains, 1 cup of each. Protein (P), Carbohydrates (C), and Fat (F), all given in grams.

	Calories	P	C	F
Barley	936	26	170	5.7
Millet	660	20	150	5.9
Rice, Brown	667	14	140	4.3
Rye Berries	591	21	130	3.9
Wheat Berries	578	24	120	4.7

Questions about a table usually ask you to find a particular value or to compare two or more values.

Example Questions

1. How many grams of carbohydrates are contained in 1 cup of rye berries?

Answers

1. The answer, **130**, is written at the intersection of the "Rye Berries" row and the C (Carbohydrates) column.

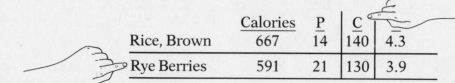

	Calories	P	C	F
Rice, Brown	667	14	140	4.3
Rye Berries	591	21	130	3.9

2. What is the *average* value of protein in 1 cup of the 5 grains shown?

2. The *average* value is found by adding the 5 protein values together and then dividing by 5.

```
  26
  20
  14
  21
+ 24
─────
 105
```

Answer: 21 grams

```
    21
5) 105
```

Answer the questions below about the following nutrition chart. Remember that the title and other parts of the chart can give you important information. Read them carefully.

Nutritional Values of Selected Meats
(Each value is for a six-ounce serving)

Meat	Calories (nearest 10)	Protein (g*) (nearest 1)	Fat (g*) (nearest 1)	Iron (mg*) (nearest .1)
BEEF				
lean ground	370	47	19	6.0
round steak	440	48	26	5.9
CHICKEN (skinned)				
light meat	280	54	6	2.2
dark meat	300	48	11	2.9
HAM (lean)	640	40	52	5.2
LAMB (lean)	480	43	32	2.9

* g = grams   mg = milligrams

1. How many milligrams of iron are contained in a six-ounce serving of round steak?

2. How many more grams of fat are contained in six ounces of chicken dark meat than in light meat?

3. Of the meats shown, what is the *median* number of calories for the six-ounce servings of meat shown?

4. What is the *ratio* of the number of calories in lean ham to the number of calories in chicken dark meat?

5. What is the *mean* number of grams of protein for the six-ounce servings of meat shown?

What Is a Graph?

A *graph* is a pictorial display of information. As a picture, a graph enables you to get a quick impression of the data being shown. It also enables you to quickly compare one number with another. On the pages ahead, we'll briefly look at the four most common types of graphs: the **circle graph**, the **bar graph**, the **pictograph**, and the **line graph**.

Reading a Circle Graph

A circle graph is drawn as a divided circle. Each segment (part) is given a name and value. The whole circle represents all (100%) of the data being displayed.

- In the most common circle graph, each segment is given a percent value. The sum of all segments adds up to 100%. See Graph A below.

- In a second type, each segment is given a number of cents as a value. Here the sum of segments adds up to $1.00. See Graph B below.

Examples:

Graph A: Johnson Family Budget

Graph B: Apene Corporation Sales, 1987 (Total sales: $250,000)

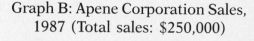

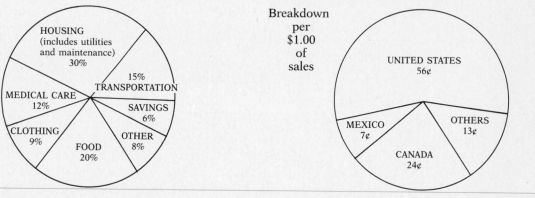

Questions about circle graphs usually ask you to find an amount of money, to find a percent, or to compare two or more values.

Example Questions

GRAPH A

1. How much more of their budget do the Johnsons spend for housing than they spend for food?

2. If the Johnson family monthly income is $1,200, what dollar amount do they pay for housing?

Answers

GRAPH A

1. The answer, **10%**, is computed by subtracting the food percent (20%) from the housing percent (30%).

2. The answer, **$360**, is obtained by computing 30% of $1,200.

164

3. What is the ratio of the Johnsons' medical care expense to their food expense?

GRAPH B

1. What percent of Apene sales during 1987 was made in Canada?

2. What dollar amount of sales was made in Canada during 1987?

3. The answer, $\frac{12}{20} = \frac{3}{5}$, is found by putting the medical care percent (12) over the food percent (20).

GRAPH B

1. See the segment labeled "Canada." 24¢ out of each sales dollar means that **24%** of all sales were made in Canada.

2. The dollar amount, **$60,000**, is obtained by computing 24% of $250,000, Apene's total sales during 1987.

Answer each question below about the following circle graph.

graph title $\longrightarrow$ PYUN FAMILY BUDGET
(monthly income: $1,500)

Breakdown per $1.00 of monthly income

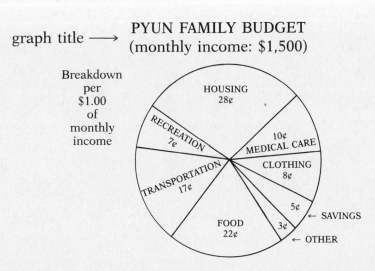

6. According to the graph title, what is the Pyun family monthly income?

8. What percent of their monthly income do the Pyuns spend on housing?

10. What is the *ratio* of the amount the Pyuns spend on recreation to the amount spent on medical care?

7. Out of each $1.00 of monthly income, how many cents do the Pyuns spend on housing?

9. What total dollar amount do the Pyuns spend on housing each month?

11. Each month the Pyuns donate $1\frac{1}{2}\%$ of their monthly income to their church. How many dollars each month is this donation?

Reading a Bar Graph

A bar graph gets its name from the thick bars that it uses to show data. These bars may be drawn either vertically (up and down) or horizontally (across).

Numerical values are read along numbered scales called *axes* that make up the sides of the graph. You read a value for each bar by finding the number on the **axis** that is across from the end of that bar. A long bar will have a greater value than a short bar.

Example: Below is a bar graph showing the growth of the population in the United States. Numerical values for the population appear on the left, along the vertical axis. The year the population is measured is written along the bottom, beneath the horizontal axis.

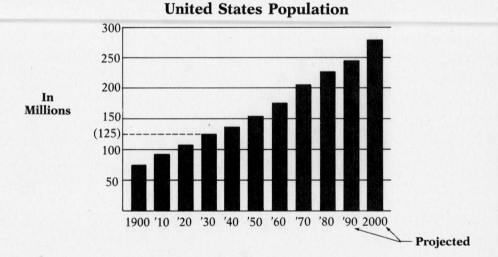

United States Population

Questions about bar graphs usually ask you to find a specific value or to compare one value with another.

Example Questions

1. What was the United States population in 1930?

2. What is the approximate ratio of the 1960 population to the 1900 population?

Answers

1. First locate the bar that is above the year 1930. Then, scan directly to the left from the top of the bar across to the vertical axis. As read on the axis, the value of the 1930 population is between 100 and 150, or **about 125 million**.

2. To compute the ratio, divide the approximate 1960 population (175) by the approximate 1900 population (75).

   Ratio $= \frac{175}{75} = \frac{7}{3}$

166

Answer the questions below the following bar graph. Choose the best answer from among the answer choices given.

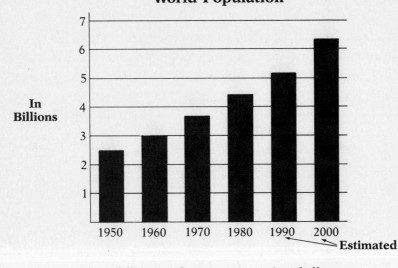

World Population

12. What was the world's population in the year 1980?

   a) 3 billion
   b) 3.6 billion
   c) 4.4 billion
   d) 5.6 billion
   e) 5.9 billion

13. By how many people is the world's population estimated to increase between the years 1980 and 2000?

   a) 1 billion
   b) 1.9 billion
   c) 2.5 billion
   d) 3.2 billion
   e) 3.6 billion

14. What is the *median* value of the population totals for the 6 specific years shown?

   a) 2.5 billion
   b) 3.0 billion
   c) 3.7 billion
   d) 4.1 billion
   e) 4.8 billion

15. What is the approximate *ratio* of the estimated population in the year 2000 to the actual population in the year 1960?

   a) 2 to 1
   b) 3 to 1
   c) 4 to 1
   d) 5 to 4
   e) 6 to 5

16. What *percent increase* in world population occurred between 1950 (2.5 billion) and 1960 (3.0 billion)?

   a) 5%
   b) $12\frac{1}{2}$%
   c) 20%
   d) 50%
   e) 75%

Reading a Pictograph

A pictograph looks similar to a bar graph. However, a pictograph uses small pictures as symbols. Each symbol has a certain value that is shown in a key written on the graph.

   To find the total value of a line of symbols, you multiply the number of symbols by the value of a single symbol. A long line of symbols has a larger value than a short line because it contains more symbols.

Example: Below is a pictograph showing the sales of bicycles in the northwest states by the Northwest Bicycle Company. The names of the states are listed along the vertical axis at left. Sales for each state are read as a number of bicycle symbols extending as lines to the right.

NORTHWEST BICYCLE COMPANY
First Quarter Bicycle Sales in Northwest States

Questions about pictographs, like bar graphs, most often ask you to find a specific value or to compare one value with another.

Example Questions

1. For the sales period shown, how many bicycles did Northwest sell in Idaho?

2. What is the ratio of sales in Oregon to the sales in Washington?

Answers

1. The first step is to count the number of bicycle symbols to the right of Idaho. Now multiply these 3 symbols by the value of each symbol, 150.

   Answer = 450 bicycles.

2. The easiest way to find this ratio is simply to put the number of Oregon symbols over the number of Washington symbols.

   Ratio $= \frac{4}{8} = \frac{1}{2}$

Answer the questions below about the following pictograph.

HIGH FLYER KITE COMPANY
Spring Season Sales Figures

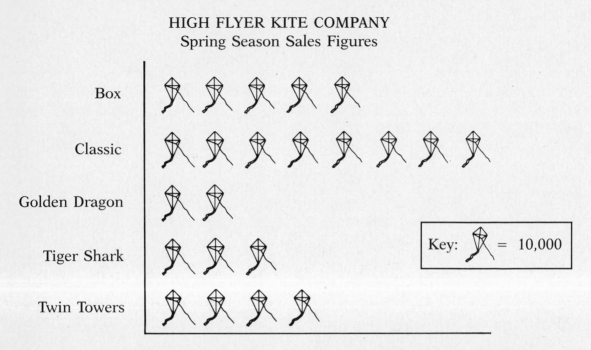

Key: = 10,000

17. What number does the symbol stand for?

18. How many Golden Dragon kites did the High Flyer Company sell during the spring season?

19. How many more Box kites were sold than Tiger Shark kites during this season?

20. What was the total number of kites sold by High Flyer for the time period shown?

21. Which style of kite is the most popular among the designs sold by High Flyer?

22. What is the **ratio** of the number of Box kites sold to the number of Classic kites sold?

23. Out of the 5 styles shown, for which kite did the **median** number of sales occur?

24. What is the **average** number of sales for the various styles of kites sold by High Flyer Company?

Reading a Line Graph

A line graph gets its name from the thin line that it uses to show data. Because every point on the line has a value, a line graph shows how data change in a continuous way. In fact, you can read as many data points as you want. You are not limited to just a few points as you are with the other graphs we've discussed.

Like a bar graph, a line graph has numbered scales called **axes**. The value of each point on the line is read as two numbers, one taken from each axis.

Example: Below is a growth chart that shows how the height of an average-size girl changes between birth and age 10. Height is read along the vertical axis, and the age at which that height occurs is read along the horizontal axis. As an example, dotted lines drawn from a point on the line indicate that an average-size 4-year-old girl has a height of about 40 inches.

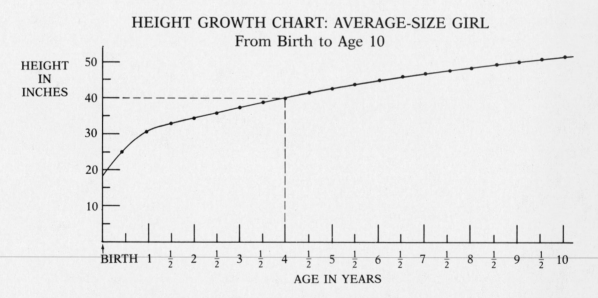

HEIGHT GROWTH CHART: AVERAGE-SIZE GIRL
From Birth to Age 10

Questions about line graphs usually are concerned with how data are changing as you move from left to right across the graph.

Example Questions

1. During which year is a girl's height increasing most rapidly?

2. What is the approximate ratio of a girl's height at age 8 to her height at age 4?

Answers

1. To answer this question, look for the year at which the graph is rising most rapidly (is the steepest).

   Answer: From birth to age 1

2. To compute this ratio, put the height at age 8 over the height at age 4.

   Answer: $\frac{50 \text{ inches}}{40 \text{ inches}}$ **or about** $\frac{5}{4}$

170

Answer the questions below about the following line graph. Choose the best answer from among the answer choices given.

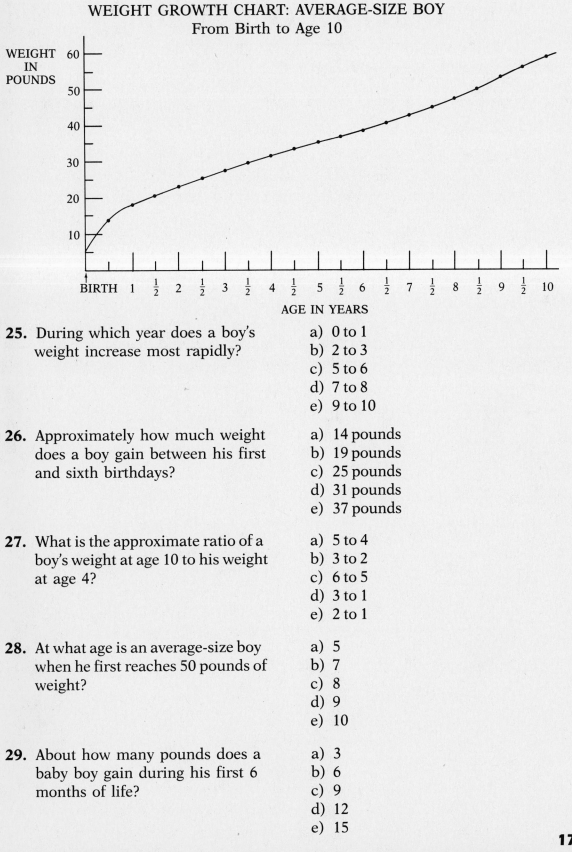

WEIGHT GROWTH CHART: AVERAGE-SIZE BOY
From Birth to Age 10

25. During which year does a boy's weight increase most rapidly?

a) 0 to 1
b) 2 to 3
c) 5 to 6
d) 7 to 8
e) 9 to 10

26. Approximately how much weight does a boy gain between his first and sixth birthdays?

a) 14 pounds
b) 19 pounds
c) 25 pounds
d) 31 pounds
e) 37 pounds

27. What is the approximate ratio of a boy's weight at age 10 to his weight at age 4?

a) 5 to 4
b) 3 to 2
c) 6 to 5
d) 3 to 1
e) 2 to 1

28. At what age is an average-size boy when he first reaches 50 pounds of weight?

a) 5
b) 7
c) 8
d) 9
e) 10

29. About how many pounds does a baby boy gain during his first 6 months of life?

a) 3
b) 6
c) 9
d) 12
e) 15

TOPIC 5:
Introduction to Probability

Chance and Outcome

Probability is the mathematical study of chance. An example can best show what the word *chance* means.

If you spin the spinner at right, you do not know where it will stop. Once you let go, the spinner is no longer in your control. The most you can say is that it is equally likely to stop in any of the four sections, A, B, C, or D. We say that "where it stops is left to chance."

Spinner

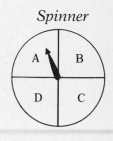

As illustrated by the spinner, we use the word *chance* to indicate our lack of control over how something will turn out. The result — where the spinner will stop — is not something that we can predict with certainty.

When we study probability, we call a result an ***outcome***. For each spin of the spinner, there are 4 possible outcomes: A, B, C, and D. Because each section is the same size, each of the 4 outcomes is equally likely to occur.

Expressing Probability as a Number

Probabilities are expressed as numbers ranging from 0 to 1, or as percents ranging from 0% to 100%.

Probabilities of 0 or 1

A probability of 0 (or 0%) means that an outcome cannot possibly occur. For example, the probability that the spinner will stop on E is 0. This is because there is no E on the circle!

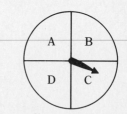

Will it stop at E??
Probability = 0 or 0%

A probability of 1 (or 100%) means that the outcome will occur for sure. In fact, a probability of 1 means that you are sure of an outcome. You can say that the probability that the spinner will stop *somewhere* in the circle is 1. There is no other possibility, as-suming that the spinner can't keep spinning forever!

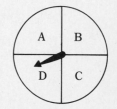

Will it stop on the circle?
(at A, B, C, or D?)
Probability = 1 or 100%

Probabilities Between 0 and 1

Almost all probabilities you'll ever think about will be between 0 and 1. For example, you might ask, "What is the probability that the spinner will land in section B?"

Answer: Because there are 4 equally likely outcomes, the chance is 1 in 4 that the spinner will land in section B.

To express the chance 1 in 4 as a probability, we write a fraction. The denominator of the fraction is equal to 4, the total number of possible outcomes.

The probability that the spinner will stop in section B is $\frac{1}{4}$. Expressed as a percent, the probability is 25%.

Probability of outcome B $= \frac{1}{4}$ or 25%

Answer each question below about probability. Express each answer both as a fraction and as a percent.

1. What is the probability that the sun will rise in the east tomorrow?

   fraction: _____ percent: _____

2. What is the probability that you will be younger tomorrow than you are today?

   fraction: _____ percent: _____

3. If you spin the spinner shown below, what is the probability that the spinner will stop in section C?

   fraction: _____ percent: _____

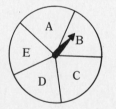

4. A die (one of a pair of dice) has 6 faces as shown. If you roll the die, what is the probability that you'll role a ⚁ ?

   fraction: _____ percent: _____

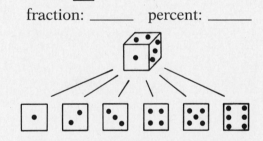

5. Suppose you shut your eyes and choose a penny from the group below. If you don't know where the "head's up" one is, what is the probability that you will choose it on your first try?

   fraction: _____ percent: _____

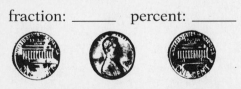

6. Below is a spinner for a money game. What is the probability that the player will win $50 on one spin?

   fraction: _____ percent: _____

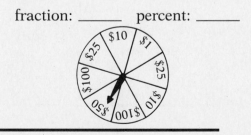

Multiple Chances for the Same Outcome

Now we want to talk about problems where an outcome can occur in more than one way. To see how this can happen, look at this next example.

Example 1: If you spin the spinner at right, what is the probability that it will stop on an A section?

First, notice that the spinner is again equally likely to stop on any one of 4 sections. But, since 2 sections are labeled A, the spinner has 2 chances in 4 of stopping on an A section.

The probability of stopping on an A section is the fraction $\frac{2}{4}$. The numerator 2 is the number of ways outcome A can occur. The denominator 4 is the total number of possible outcomes — the total number of sections in the circle.

$$\text{Probability of an outcome A} = \frac{2}{4} \text{ (which can be reduced)}$$
$$= \frac{1}{2} \textbf{ or 50\%}$$

Example 1 helps you understand the mathematical definition of probability:

$$\text{Probability of an outcome} = \frac{\text{number of ways an outcome can occur}}{\text{total number of possible outcomes}}$$

Example 2: Below are 6 cards placed on a table. Two are aces and 4 are kings. If you close your eyes, shuffle the cards around, and then pick a card without looking, what is the probability that you'll pick a king?

Since there are 6 cards, the total number of possible outcomes is 6. Therefore, 6 is the denominator of the probability fraction.

To determine the numerator, notice that you can choose any one of 4 cards and get a king. This means there are 4 ways that the **king outcome** can occur. Thus, the numerator is 4.

Probability of picking a king $= \frac{4}{6} = \frac{2}{3}$ **or $66\frac{2}{3}$%**

See if you can figure out the probability of picking an ace on your first try.

Answer each question below. Express each answer as a fraction and as a percent.

7. If you spin the spinner below, what is the probability of an outcome C?

fraction: _____ percent: _____

8. If you randomly choose from the cards below, what is the probability that you'll choose a "2" on your first try?

fraction: _____ percent: _____

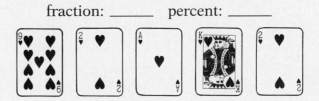

9. Suppose, with your eyes closed, you pick a single penny from those shown below. What is the probability that you'll pick a "tail's up" penny?

fraction: _____ percent: _____

10. For a Christmas party, the names of 10 people are "put in a hat." Each person then draws the name of a friend for whom to buy a gift. Four women and 6 men are in the group. What is the probability that the first person who draws will get a woman's name?

fraction: _____ percent: _____

11. If you randomly open a 200-page book to any page, what is the probability that you'll be on a page whose page number ends in the digit 0?

fraction: _____ percent: _____

12. If you randomly pick a number from 1 to 25, what is the probability that the number you pick will be evenly divisible by 3?

fraction: _____ percent: _____

13. On one roulette wheel, there are 24 numbers colored red, 24 numbers colored black, and 2 numbers colored green. On a single bet, what is the probability of *losing* if you choose both green numbers? (Hint: The probability of losing is the probability that something other than one of the 2 green numbers will be chosen.)

fraction: _____ percent: _____

14. If you randomly choose a card from the cards below, what is the probability that you *won't choose* a face card?

fraction: _____ percent: _____

Dependent Probabilities

For these last two pages of probability, we want to look briefly at the topic of **dependent probabilities**. The following example will show what we mean.

Example: Let's look again at the 6 cards placed on the table. Two are aces and 4 are kings. Again pretend you close your eyes and pick a card. Only this time, pretend you look at it, put it aside, and then pick a second card.

a) What is the probability of picking an ace for your first card?

b) What is the probability of picking a king for your second card?

- The probability of picking an ace for your first card is $\frac{2}{6} = \frac{1}{3}$. This is because there are 2 aces out of a total of 6 cards.

- The answer to question b *depends on what card you actually get as your first card.*

Because the answer to b depends on what card you pick first, the answer to b is called a **dependent probability**.

Let's look at the two possible answers to b.

Possibility 1: Your first card is a king.

remaining cards:

 After you pick a card, there are only 5 cards left. If your first card is a king, 3 of the remaining 5 cards are kings and 2 are aces.

In this case, **the probability that you'll choose a king for your second card is $\frac{3}{5}$ or 60%.**

Possibility 2: Your first card is an ace.

remaining cards:

 If the first card chosen is an ace, 4 of the remaining 5 cards are kings and 1 is an ace.

In this case, **the probability that you'll choose a king for your second card is $\frac{4}{5}$ or 80%.**

Answer each question below. Express each answer as a fraction and as a percent.

15. Suppose you randomly choose 2 cards from the group of cards shown at the right. What is the probability of choosing a face card as your second card if your first card turns out to be a face card?

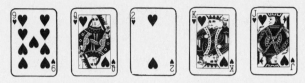

fraction: _____ percent: _____

16. During the holidays, 4 friends agreed to exchange gifts. Each person wrote his or her name on a slip of paper. Then each randomly chose the name of a friend. There are 3 women and 1 man in the group.

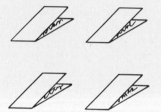

a) What is the probability that the first person who draws will draw a woman's name?

fraction: _____ percent: _____

b) If the first person does draw a woman's name, what is the probability that the second person who draws will also draw a woman's name?

fraction: _____ percent: _____

17. Lila bought 3 cans of corn and 3 cans of peas. Except for the labels, the cans look identical. As she walked in the door at home, she heard the phone ring. She placed the bag with the cans on the floor and dashed into the living room to the phone. While she talked, Lila's three-year-old son Jeffy peeled the labels off all 6 cans!

a) When she opens a can, what is the probability that Lila will open a can of corn?

fraction: _____ percent: _____

b) If the first can she opens is peas, what is the probability that the second can she opens will also be peas?

fraction: _____ percent: _____

c) If the first 3 cans Lila opens are peas, what is the probability that the next can she opens will be corn?

fraction: _____ percent: _____

Pre-GED Math Book II Posttest

The following set of 40 questions will give you a chance to briefly review many of the skills you have developed in this workbook. The questions are designed to be much like those you'll see on the GED and other math tests. For each question, you choose an answer from the choices given.

Work carefully and answer every question. When you finish, check your answers with the answers given in the answer key on pages 203-204.

---

1. When gas is selling for $1.068 per gallon, how much would you actually have to pay for a total purchase of 1 gallon?

   a) $1.00 b) $1.06 c) $1.07 d) $1.08
   e) $1.09

2. The distance between Tami's office and the library is about .8 miles. From the library to the post office is 1.4 miles. Passing the library on the way, how many miles is Tami's office from the post office?

   a) .6 b) 1.2 c) 1.8 d) 2.2 e) 2.8

3. At Janet's Shop for Women, Janet sells Smart Set skirts for $32.48. If Janet is able to buy these skirts from the manufacturer for $18.99 each, how much profit does she make on each skirt she sells?

   a) $12.47 b) $13.01 c) $13.49 d) $13.97
   e) $14.01

4. Bernie purchased an electric drill on sale for $39.99. He also bought a set of drill bits for $16.49. How much change will Bernie receive if he pays for these items with three $20 bills?

   a) $1.32 b) $1.98 c) $2.92 d) $3.52
   e) $4.02

5. Earning $5.65 per hour as a carpenter's assistant, how much does Jason earn in 35 hours of work each week?

   a) $127.25 b) $168.45 c) $197.75
   d) $209.25 e) $217.75

6. Brandi paid $5.23 for 3.6 pounds of chicken. At this rate, how much is she paying for each pound? Express your answer to the nearest penny.

   a) $1.41 b) $1.45 c) $1.53 d) $17.69
   e) $18.79

7. For exercise, Roger walks around Fenton Park on each of three days during the week. On each of the other four days he walks 1.5 miles around the neighborhood grade school. If the distance around Fenton Park is 3.2 miles, how many total miles does Roger walk for exercise each week?

   a) 4.7 b) 9.6 c) 11.4 d) 15.6 e) 25.2

8. Walking at a rate of 3.4 miles per hour, how long would it take Debbie to walk from her house to the library and back if the library is 5 miles from her house? Express your answer to the nearest tenth of an hour.

   a) 1.4 b) 1.5 c) 1.6 d) 2.8 e) 2.9

9. Opal and 3 friends agreed to split the cost of lunch. Each agreed to pay an equal amount. The bill included the items listed below. From this bill, determine Opal's share of the bill.

Sandwiches: $13.78, Salads: $ 6.04, Drinks: $ 3.74, Desserts: $ 7.08

a) $7.66 b) $7.96 c) $8.16 d) $8.36
e) $8.76

10. At a January sale, Gordon bought a microwave oven for $379. He made a down payment of $59. He then agreed to pay off the balance in 3 equal monthly payments. If he was charged no interest, which arithmetic expression correctly shows the amount of each of Gordon's monthly payments?

a) $379 − $59
b) ($379 + $59) ÷ 3
c) $379 − ($59 ÷ 3)
d) ($379 − $59) ÷ 3
e) $379 ÷ 3 − $59

11. There are 52 weeks in a year. What fraction of a year is 36 weeks?

a) $\frac{9}{52}$ b) $\frac{3}{5}$ c) $\frac{9}{13}$ d) $\frac{13}{9}$ e) $\frac{26}{9}$

12. For Halloween, Lorri bought 48 bags of candy and a dozen apples to give to children. By 8:00 that evening, she had already given away 36 bags of candy. At that time, what fraction of the original amount of candy did she still have left?

a) $\frac{1}{4}$ b) $\frac{1}{3}$ c) $\frac{1}{2}$ d) $\frac{2}{3}$ e) $\frac{3}{4}$

13. Leon invited several friends over for dinner on Tuesday evening. For that dinner he prepared 3 steaks. One weighed $\frac{5}{8}$ pound, the second weighed $1\frac{1}{2}$ pounds, and the third weighed $\frac{3}{4}$ pound. How many pounds of steak did Leon prepare?

a) $1\frac{5}{8}$ b) $1\frac{9}{14}$ c) $1\frac{7}{8}$ d) $2\frac{7}{8}$ e) $3\frac{7}{8}$

14. During the first week of his diet, Ernie lost a total of $1\frac{5}{8}$ pounds. During the second week, he lost only $\frac{9}{16}$ pounds. How much more did Ernie lose the first week than the second week? Express your answer in pounds.

a) $\frac{1}{16}$ b) $\frac{9}{16}$ c) $1\frac{1}{16}$ d) $2\frac{1}{16}$ e) $2\frac{3}{16}$

15. Marian spends $\frac{1}{4}$ of her monthly income of $824 for rent. She also spends $\frac{1}{8}$ of her income on food. Knowing this, determine what amount Marian pays for rent each month.

a) $103 b) $206 c) $250 d) $424 e) $618

16. To make light green paint, the directions say to mix $\frac{3}{8}$ quart of emerald green with each gallon of white. If Mark has only $\frac{2}{3}$ gallon of white, what fraction of a quart of emerald green should he use to make the light green color?

a) $\frac{1}{4}$ b) $\frac{1}{3}$ c) $\frac{7}{24}$ d) $\frac{9}{24}$ e) $\frac{1}{2}$

17. For the 4th of July picnic, Teresa is preparing a selection of meat dishes. She is cooking $6\frac{1}{2}$ pounds of hot dogs, $5\frac{3}{4}$ pounds of hamburger, and $4\frac{5}{8}$ pounds of cold cuts. Assuming that each person will eat $\frac{5}{8}$ pounds of meat during the day-long picnic, how many people will Teresa's meat dishes serve?

a) 13 b) 19 c) 24 d) 27 e) 31

18. Gail is lining up packages according to weight. She places the heaviest package at the left and the lightest package at the right. Which listing below shows the correct order in which Gail should line up the 4 packages labeled below?

Package	Weight
A	$\frac{7}{8}$ pound
B	.8 pound
C	$\frac{5}{7}$ pound
D	.79 pound

a) B, C, A, D b) A, C, B, D c) C, D, B, A
d) A, C, D, B e) A, B, D, C

19. Norm cut a 17-foot-long "two-by-four" into 9 pieces as follows: First, he cut off a piece that measured $5\frac{5}{6}$ feet long. Next, he cut off a piece that measured $6\frac{1}{2}$ feet long. Last, he cut the remaining piece into 7 equal pieces. Assuming no waste as he cut, how many inches long are each of these 7 short pieces?

a) 7 b) 8 c) 9 d) 10 e) 11

20. For use in his auto shop, Curt bought $\frac{3}{4}$ pound of lock washers and $\frac{5}{8}$ pound of assorted washers. Each type of washer costs $1.29 per pound. He paid for his purchase with a $20 bill. Which arithmetic expression below correctly shows the amount of change he should receive?

a) $20 + $1.29 $(\frac{3}{4} + \frac{5}{8})$

b) $20 − $1.29 ÷ $(\frac{3}{4} + \frac{5}{8})$

c) $20 − $1.29 $(\frac{3}{4} + \frac{5}{8})$

d) $20 − $\frac{3}{4} + \frac{5}{8}$

e) $1.29 $(\frac{3}{4} + \frac{5}{8})$ − $20

21. By the time he went to bed Saturday night, Alan had completed 30% of the work he hoped to get done that weekend. What fraction of the work had he completed by that evening?

a) $\frac{3}{100}$ b) $\frac{3}{10}$ c) $\frac{1}{3}$ d) $\frac{3}{7}$ e) $\frac{3}{4}$

22. From goal line to goal line, a football field is 100 yards long. When Keith was in the eighth grade, he could stand on one goal line and throw a football to the 39-yard line. What percent of the length of the football field could Keith throw a football at that time?

a) $\frac{39}{100}$% b) .39% c) 39% d) 61%

e) 139%

23. At the end of each month Vera is given a 4% bonus based on the dollar value of sales she makes. If she sold $2,658 worth of clothes during July, how much would Vera's bonus be for that month?

a) $47.82 b) $66.45 c) $98.72
d) $106.32 e) $2,764.32

24. One mile is equal to 5,280 feet. What percent of a mile is 440 yards?

a) 12.5% b) 25% c) $33\frac{1}{3}$% d) 37.5%
e) 66%

25. A pair of wool pants at Tom's Men's Store is priced at $38.50. How much would you expect to pay for these pants in a state with a 6% state sales tax?

a) $2.61 b) $40.81 c) $41.11 d) $44.70
e) $64.17

26. Before Leslie was laid off from her job at the auto factory, she earned $11.25 per hour. On her new job as a paint salesperson, she earns $7.50 per hour. What percent salary decrease has Leslie taken in accepting this new job?

a) $16\frac{2}{3}$% b) 25% c) 30% d) $33\frac{1}{3}$%

e) $37\frac{1}{2}$%

27. During a time of high inflation, the cost of food products rose 8% between January and September. At this rate, what would you expect the price of a loaf of bread to be in September if it sold for $1.25 in January?

a) $.10 b) $1.15 c) $1.29 d) $1.33
e) $1.35

28. On the local news it was reported that 17% of the citizens of Portland who were of voting age favored lowering the speed limit within the city limits. The newscaster said that 35,190 people would favor this change. From this information, figure out how many people of voting age live in Portland.

a) 207,000 b) 227,000 c) 267,000
d) 307,000 e) 351,900

29. Chun paid a sale price of $31.85 for a bedspread that was marked "35% off original price" and $63.70 for a quilt. What would she have paid for the bedspread if it hadn't been on sale?

a) $20.70 b) $42.00 c) $49.00 d) $66.85
e) $147.00

31. Using the drawing, determine the difference in length between the two sizes of screws shown. Express your answer to the nearest sixteenth of an inch.

a) $\frac{5}{16}$ b) $\frac{7}{16}$ c) $\frac{9}{16}$ d) $\frac{11}{16}$ e) $\frac{13}{16}$

1 inch:

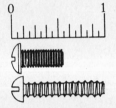

33. Using the simple interest formula I = PRT, determine the amount of interest you'd pay on a loan of $3,000 borrowed for 2 years 8 months at an interest rate of 12%.

a) $90.80 b) $96.00 c) $800.00
d) $908.00 e) $960.00

30. As an appliance salesperson, Brett is paid a monthly salary of $620. He also receives a 2% commission on the dollar value of sales that he makes during the month. Which arithmetic expression correctly shows the amount of Brett's monthly pay during a month in which he makes $3,980 in sales?

a) .02 × $3,980
b) .02 ($620 + $3,980)
c) (.02 × $3,980) − $620
d) (.02 × $3,980) + $620
e) (.02 × $620) + $3,980

32. When he put a new starter on his Japanese sports car, Adam noticed that the new starter was 6 millimeters longer than the old starter. If the old starter is 19.2 centimeters long, how many centimeters long is the new starter?

a) 19.26 b) 19.6 c) 19.62 d) 19.8 e) 25.2

34. In Beth's Monday night GED class there are five women students and six men. Looking at the list of their ages shown below, figure out the *mean* (average) age of these women students.

Mary   18
Joyce   35
Amy   37
Helen   52
Sarah   68

a) 37 b) 42 c) 52 d) 56 e) 59

35. To make a fruit salad for the school picnic, Sonja mixed 6 cups of chopped apples, 4 cups of sliced bananas, $1\frac{1}{2}$ cups of blueberries, $2\frac{1}{2}$ cups of strawberries, and 4 cups of oranges. In this salad, what is the *ratio* of oranges to apples?

a) $\frac{2}{3}$ b) $\frac{6}{5}$ c) $\frac{3}{2}$ d) $\frac{1}{2}$ e) $\frac{1}{4}$

36. Suppose you randomly choose two cards from those shown below. If your first card turns out to be a face card, what is the probability that your second card will be a 4?

a) 5% b) 13% c) 20% d) 25% e) 75%

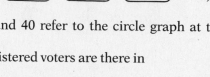

37. Walter bought a box of 6 popsicles for his 3 children. The box contained 2 orange popsicles, 2 grape, and 2 cherry. The problem was that each child wanted grape! So, to be fair, Walter had each child take a popsicle out of the box without first seeing the color. If the first child pulled out an orange one, what is the probability that the second child will be luckier and pull out a grape popsicle?

a) $\frac{1}{6}$ b) $\frac{1}{5}$ c) $\frac{1}{3}$ d) $\frac{2}{5}$ e) $\frac{3}{5}$

Questions 38, 39, and 40 refer to the circle graph at the right.

38. How many registered voters are there in Linn County?

a) 26,500 b) 29,150 c) 55,650
d) 66,250 e) 100,000

Registered Voters Of Linn County

39. Determine what percent of the registered voters in Linn County are registered as Democrats.

a) 28% b) 40% c) 47% d) 51% e) 58%

40. What percent of the registered voters in Linn County are not registered as either Democrats or Republicans?

a) 4% b) 7% c) 9% d) 12% e) 16%

Democrats
(26,500)

Other
(10,600)

Republicans
(29,150)